Lecture Notes in Geoinformation and Cartography

Series Editors: William Cartwright, Georg Gartner, Liqiu Meng, Michael P. Peterson

For further volumes:
http://www.springer.com/series/7418

Erwan Bocher • Markus Neteler
Editors

Geospatial Free and Open Source Software in the 21st Century

Proceedings of the first Open Source Geospatial Research Symposium, OGRS 2009

Springer

Editors
Erwan Bocher
Institute for Research on Urban Sciences
and Techniques (IRSTV) FR CNRS 2488
Ecole Centrale de Nantes
1 rue de la Noé
44321 Nantes Cedex 3

Markus Neteler
Fondazione Edmund Mach
Via E. Mach 1
38010 San Michele all'Adige
Italy

ISSN 1863-2246 e-ISSN 1863-2351
ISBN 978-3-642-43361-0 ISBN 978-3-642-10595-1 (eBook)
DOI 10.1007/978-3-642-10595-1
Springer Heidelberg Dordrecht London New York

Softcover reprint of the hardcover 1st edition 2012

Printed on acid-free paper

Springer is part of Springer Science+Business Media (www.springer.com)

Preface

Methods, software and tools in the geospatial field have been strongly growing for 20 years. The ongoing digitalization with data exchange across often very heterogeneous and physically disconnected groups requires the availability and implementation of data exchange standards. Especially Open Source GIS keep well pace with the adoption and sometimes even definition of standards; the software development is furthermore pushed by the raise of new standards within participatory and exchange platforms. Other needs are driven by territorial decision makers guiding towards sustainable development and data sharing. Decision makers and leaders are greatly supported through the availability of Geospatial Free and Open Source Software (GFOSS) which includes geospatial programming libraries, desktop GIS, web mapping, and catalog systems for handling metadata. The GFOSS history lasts back more than 25 years when the first desktop GIS, GRASS became available with its numerous scientific applications (landscape ecology, epidemiology, remote sensing, urban planning, 3D representations). With the advent of wider Internet usage, the researchers at the University of Minnesota gave birth to MapServer, the first Open Source platform for publishing geospatial data and interactive mapping applications in the Web. As of today, more than 350 GFOSS tools are listed in the FreeGIS web site (http://www.freegis.org). With no doubt, GFOSS has created a new dynamic in the GIS field, largely promoted by the Open Source Geospatial Foundation (http://www.osgeo. org) and its series of Free Open Source Software for Geospatial conferences (FOSS4G).

Many applications have their roots in the heart of research activities and likely this will remain in future. In this context, the initial goal of the International Open Source Geospatial Research Symposium (OGRS) 2009 was to provide a large panel of the innovative works currently carried out within research projects around the world. And, what renders a symposium of this shape particularly interesting is the focus on Open Source development: an entire research and innovation symposium dedicated to free and open source geospatial methods and technologies. The first symposium was held in Nantes (France) from July 8–10, 2009, and organized by IRTSV (Research Institute on Urban Sciences and Techniques), a Research Federation founded by CNRS (*Centre National de la Recherche Scientifique*) and the French Ministry of Research.

The organizers of OGRS wanted to stand up for the idea that leading research projects with open source software as backbone will help to offer better solutions for the manifold challenges which our society is facing today, especially in the field of sustainable development. During the last few years, new communities including researchers, local authorities GIS administrators, companies and citizens have appeared. This evolution has lead to new dynamics in the geomatics sphere towards finding solutions together by sharing knowledge and methods without starting with license constraints. In this framework, OGRS shall represent a hub to network, share and develop ideas for future collaborative work between academic research activities and everyday operational activities (companies, local authorities, and others).

This book aims at unveiling new ideas and methods provided in the field of research in 2009 in this symposium. It presents a collection of full scientific papers accepted at the first OGRS symposium. Additional best short papers which had been selected by the scientific committee were extended and also published in this book. We have included not only original scientific research papers, but also selected papers on applied GIS-related issues (feedback and experience, concepts, software development, etc.) which were also important topics for the symposium's aims.

The articles are organized in four chapters that intend to summarize recent progress and identify key research issues concerning: geoprocessing, remote sensing, cartography and standards, and geospatial web. The last chapter is especially dedicated to feedback and future.

To make GFOSS activities and their outstanding results better understandable to a wide audience, a series of chapters were derived from selected conference contributions. The scope of this book is to illustrate the variety of functionalities which GFOSS software covers today as well as the high level of expertise being the framework for such development.

We hope that our readers will find the chapters included in this book interesting and we believe that they may represent a helpful reference in the future for all those who work in any of the GFOSS research areas.

Erwan Bocher and Markus Neteler

Nantes
December 2010

Programme Committee

Programme Chair Erwan Bocher,
IRSTV – FR CNRS 2488 Ecole Centrale de Nantes (France)

Programme Co-Chair Markus Neteler, Fondazione Mach – Centro Ricerca e Innovazione, S. Michele all'Adige (Italy)

Scientific Committee

Thierry Badard, Département des sciences géomatiques, Université Laval (Quebec)
Salvador Bayarri, IVER (Spain)
Bénédicte Bucher, Laboratoire COGIT IGN (France)
Gilberto Câmara, Brazil's National Institute for Space Research (Brazil)
Christophe Claramunt, Ecole Navale de Brest (France)
Jean-Christophe Desconnet, Unité ESPACE Montpellier (France)
Olivier Ertz, Haute Ecole d'Ingénierie et de Gestion du Canton de Vaud (HEIG-VD) (Switzerland)
Ari Jolma, Research Group on Environmental Informatics and Software, Helsinki University of Technology (Finland)
Thomas Leduc, Laboratoire CERMA UMR CNRS 1563 (France)
Xiang Li, Department of Geography, East China Normal University (China)
Nathalie Long, Laboratoire LIENSs UMR CNRS 6250 Université de La Rochelle (France)
Guillaume Moreau, Laboratoire CERMA UMR CNRS 1563 (France)
Stefan Steiniger, University of Calgary, Alberta (Canada)
F. Javier Zarazaga-Soria, University of Zaragoza (Spain)

Contributing Authors

Aguiar De, Rafael Alves
Rio Janeiro State University
Brazil

Arazy Ofer
CEOS University of Alberta
Canada

Baranski Bastian
Institute for Geoinformatics
Münster
Germany

Bocher Erwan
IRSTV Ecole Centrale de
Nantes
Nantes
France

Branger Flora
Cemagref UR HHLY
Lyon
France

Brasebin Mickaël
IGN COGIT
Saint-Mandé
France

Braud Agnès
LSIIT Université de Strasbourg
France

Braud Isabelle
Cemagref UR HHLY
Lyon
France

Brauner Johannes
Technische Universität
Dresden
Germany

Brito Jorge Luís Nunese Silva
Rio Janeiro State University
Brazil

Buard Elodie
IGN COGIT
Saint-Mandé
France

Bucher Bénédicte
IGN COGIT
Saint-Mandé
France

Câmara Gilberto
INPE
São José dos Campos
Brazil

De Souza Ricardo Cartaxo Modesto
INPE
São José dos Campos
Brazil

Corbelle Rico Eduardo
Laboratorio do Territorio
Escola Politécnica superior
University Santiago de
Compostela
Spain

Crecente Maseda Rafael
Laboratorio do Territorio
Escola Politécnica superior
University Santiago de Compostela
Spain

Croituru Arie
CEOS University of Alberta
Canada

Da Silva Badolato Irving
Rio Janeiro State University
Brazil

Da Costa Sarina Lustosa
Rio Janeiro State University
Brazil

Debionne Samuel
Hydrowide
Saint-Martin d'Hères
France

Ertz Olivier
IICT, HEIG-VD,
Yverdon-les-Bains
Switzerland

Etcheverry Patrick
LIUPPA IUT de Bayonne
France

Farias Reolon Patricia
Rio Janeiro State University
Brazil

Filho Orlando Bernardo
Rio Janeiro State University
Brazil

Foerster Theodor
ITC
Enschede
The Netherlands

Follin Jean-Michel
L2G ESGT
Le Mans
France

Gaio Mauro
LIUPPA Université de Pau et des Pays de l'Adour
Pau
France

Grosso Eric
IGN COGIT
Saint-Mandé
France

Herteaux Vincent
Geomatys
Arles
France

Hunter Andrew J. S.
University of Calgary
Canada

Jankowfsky Sonja
Cemagref
Lyon
France

Joliveau Thierry
CRENAM-ISIG Université Jean-Monnet
Saint-Etienne
France

Lachiche Nicolas
LSIIT Université de Strasbourg
France

Lagarde Pierre
BRGM
Orléans
France

Laurent Maxence
IICT HEIG-VD
Yverdon-les-Bains
Switzerland

Le Ny Pierre-André
Markina Corpus
Paris
France

Lesbegueries Julien
LSIIT Université de Strasbourg
France

Loustau Pierre
LIUPPA Université de Pau et des Pays de l'Adour
Pau
France

Luong Nhan
LIUPPA IUT de Bayonne
France

Marquesuzaà Christophe
LIUPPA IUT de Bayonne
France

Martin Jean-Yves
IRSTV Ecole Centrale de Nantes
France

Mota Guilherme Lucio Abelha
Rio Janeiro State University
Brazil

Mustière Sébastien
IGN COGIT
Saint-Mandé
France

Nodenot Thierry
LIUPPA IUT de Bayonne
France

Perret Julien
IGN COGIT
Saint-Mandé
France

Puissant Anne
LIVE Université de Strasbourg
France

Randiranandrivelo Tiana
BRGM
Orléans
France

Rappo Daniel
IICT HEIG-VD
Yverdon-les-Bains
Switzerland

Ribeiro João Araújo
Rio Janeiro State University
Brazil

Royer Albert
LIUPPA Université de Pau et des Pays de l'Adour
Pau
France

Sallaberry Christian
LIUPPA Université de Pau et des Pays de l'Adour
Pau
France

Schaeffer Bastian
Institute for Geoinformatics
Münster Germany

Simonetto Elisabeth
L2G ESGT
Le Mans
France

Skupinski Grzegorz
LIVE Université de Strasbourg
France

Steiniger Stefan
University of Calgary
Canada

Taillard Eric
MIS HEIG-VD
Yverdon-les-Bains
Switzerland

Teixeira Silveira Marcelo
Rio Janeiro State University
Brazil

Vannier Olivier
LTHE Université de Grenoble
France

Viallet Pierre
Hydrowide
Saint-Martin d'Hères
France

Vinhas Lúbia
INPE
São José dos Campos
Brazil

Table of Contents

Part 1: Geoprocessing

1 A Service-Oriented Framework for Real-Time and Distributed Geoprocessing 3
Bastian Schaeffer, Bastian Baranski, Theodor Foerster, and Johannes Brauner

2 GeOxygene: Built on Top of the Expertise of the French NMA to Host and Share Advanced GI Science Research Results 21
Bénédicte Bucher, Mickaël Brasebin, Elodie Buard, Eric Grosso, Sébastien Mustière, and Julien Perret

3 Use of Open-Source GIS for the Pre-processing of Distributed Hydrological Models 35
Flora Branger, Sonja Jankowfsky, Olivier Vannier, Pierre Viallet, Samuel Debionne, and Isabelle Braud

4 A Platform for Spatial Data Labeling in an Urban Context 49
Julien Lesbegueries, Nicolas Lachiche, Agnès Braud, Grzegorz Skupinski, Anne Puissant, and Julien Perret

5 TAnaTo2: A Tool to Evaluate the Impact of Natural and Anthropogenic Artefacts with a TIN-Based Model 63
Erwan Bocher, and Jean-Yves Martin

Part 2: Remote Sensing

6 The E-Foto Project and the Research to Implement a GNU/GPL Open source Educational Digital Photogrammetric Workstation 89
Guilherme Lucio Abelha Mota, Jorge Luís Nunes e Silva Brito, João Araújo Ribeiro, Orlando Bernardo Filho, Marcelo Teixeira Silveira, Rafael Alves de Aguiar, Irving da Silva Badolato, Sarina Lustosa da Costa, and Patrícia Farias Reolon

7 An Overview on Interferometric SAR Software and a Comparison Between DORIS and SARSCAPE Packages 107
Elisabeth Simonetto, and Jean-Michel Follin

8 An Object-Oriented Approach to Automatic Classification of Panchromatic Aerial Photographs with GRASS GIS and R 123
Eduardo Corbelle Rico, and Rafael Crecente Maseda

Part 3: Cartography and Standards

9 Sensor Web Enablement Standards for Groundwater Monitoring 141
Tiana Randrianarivelo, Pierre Lagarde, and Vincent Heurteaux

10 Towards an Automated Label Placement Service 157
Olivier Ertz, Maxence Laurent, Daniel Rappo, and Eric Taillard

Part 4: Geospatial Web

11 Building Location-Based Mass Collaboration Systems: Challenges and Opportunities 175
Arie Croitoru, and Ofer Arazy

12 GeoStream: Spatial Information Indexing Within Textual Documents Supported by a Dynamically Parameterized Web Service 191
Albert Royer, Christian Sallaberry, Pierre Loustau, Mauro Gaio, Thierry Joliveau, and Pierre-André Le Ny

13 WIND: an Interaction Lightweight Programming Model for Geographical Web Applications 211
The Nhan Luong, Patrick Etcheverry, Thierry Nodenot, and Christophe Marquesuzaà

Part 5: From Feedback to Future

14 Free and Open Source GIS: Will There Ever Be a Geo-Linux? ... 229
Gilberto Câmara, Lúbia Vinhas, and Ricardo Cartaxo Modesto de Souza

15 Free and Open Source GIS Software for Building a Spatial Data Infrastructure 247
Stefan Steiniger, and Andrew J. S. Hunter

Part 1: Geoprocessing

Chapter 1 A Service-Oriented Framework for Real-Time and Distributed Geoprocessing

Bastian Schaeffer[1], Bastian Baranski[1], Theodor Foerster[2], Johannes Brauner[3]

[1]Institute for Geoinformatics, Münster, Germany
{schaeffer,baranski}@uni-muenster.de
[2]International Institute for Geo-information Science and Earth Observation (ITC), Enschede, The Netherlands foerster@itc.nl
[3]Technische Universität Dresden, Dresden, Germany
johannes.brauner@tu-dresden.de

Abstract. With increasing availability of network bandwidth, server side processing capacity and advancements regarding the standardization of Web Service technologies, it becomes feasible to process geodata over the Web. As a result, web-based geoinformation is generated, which supports in particular spatial decision making. However, existing Spatial Data Infrastructures (SDI) mainly address data retrieval and portrayal, but to provide web-based geoinformation through SDIs web-based geoprocessing is required. This was the starting point to develop a framework which integrates existing desktop Geographical Information System functionality in SDIs. Moreover, an approach will be introduced to improve complex and large-scale geoprocesses in two dimensions: On the vertical scale, by significantly increasing the performance of single processes by means of Grid Computing technology and by improving other Quality of Service aspects. On the horizontal scale, by chaining this highly efficient processing services in order to automate these business processes. The relevance of this work will be demonstrated in a real-world scenario focusing on air quality assessment based on real-time data with the help of open source components.

E. Bocher and M. Neteler (eds.), *Geospatial Free and Open Source Software in the 21st Century*, Lecture Notes in Geoinformation and Cartography,
DOI 10.1007/978-3-642-10595-1_1,

1.1 Introduction

With the advent of the Service-Oriented Architecture (SOA) paradigm (Erl 2005) and advancements in Web Services technology (Weerawarana et al. 2006), Geographical Information Systems (GIS) underwent a substantial change from stand-alone applications to distributed service architectures manifested in Spatial Data Infrastructures (SDI) (Masser 2005). While existing SDIs mainly address geodata retrieval and portrayal (Kiehle et al. 2007) the gap between available geodata and required web-based geoinformation increases, especially in the context of increased network and processing capacities.

To overcome this gap, it is a common approach in mainstream IT to generate Web Services from existing functionality by adapting and wrapping it (Papazoglou and van den Heuvel 2007). This approach is also promising for geospatial applications. It allows the service provider to integrate stand-alone functionality, such as provided by GRASS GIS,[1] into an SDI. Additionally, the integration of existing functionality is sustainable because it prevents reimplementation. Thus, geoprocesses can be exposed through distributed Web Services. Since interoperability is a key requirement of the SOA paradigm, which is ensured through standardization, open standards such as developed by the Open Geospatial Consortium (OGC) for geodata retrieval and portrayal create a strong foundation to establish Geoprocessing Services in existing SOAs.

The OGC Web Processing Service (WPS) interface specification[2] became an official standard in mid 2007 and is a major attempt to publish and perform geoprocesses on the web. Such a process can range from a simple geometric calculation (for example a simple spatial intersect operation) to a complex simulation process (for example calculating rain water discharge).

However, highly specialized geospatial applications based on large volumes of distributed data such as live sensor data streams at different scales combined with high resolution geodata, which have to be analyzed in real-time for risk management issues, require often the functionality of multiple processes. To improve the computational performance of processing such large amounts of real-time geodata, Grid Computing provides appropriate tools. Additionally, Cloud Computing (for example realized by Grid Computing) is a promising approach to improve the service

[1] GRASS GIS: http://grass.osgeo.org

[2] OpenGIS Web Processing Service Standard: http://portal.opengeospatial.org/files/?artifact_id=24151

availability and to support a specific level of Quality of Service (QoS). Although the application of Grid Computing is not novel to the mainstream IT-world (as shown e.g. by Foster and Kesselman 1998), in the context of geospatial applications and OGC Web Services only little research has been conducted; see Baranski (2008), Lanig et al. (2008) and OWS-6 (2009).

Applying Grid Computing improves the performance of a complex application only on a vertical scale, but on a horizontal scale whole business processes can be optimized by composing geoprocessing workflows incorporating different types of distributed services (for retrieval and processing) (Alameh 2003). This enables outsourcing of specific tasks to highly specialized Web Services. As a result of this approach, workflows become flexible and are able to quickly adopt to changing requirements. Therefore, cross-enterprise workflows can be composed as virtual business processes (van der Aalst and van Hee 2004) since they integrate several processes from different partners. These composed workflows appear as a single process from an external point of view. Outsourcing of specific tasks to specialized services is still missing for geoprocessing workflows (Alameh 2003). The combination of the two aspects (Grid Computing and workflow) leads to a framework for high-performance and fully automated value-added geoprocessing workflows.

The introduced framework is applied to a real-world scenario for real-time and distributed geoprocessing involving live sensor data and high-resolution geodata, both served through standardized Web Services. In the scenario a geoprocessing workflow has to be modelled to identify areas with a high Particulate Matter pollution. To accomplish this scenario, computationally extensive processes are required. The required steps in the scenario such as modelling, execution of geoprocesses and portrayal of the process results will be achieved through an integrated client application, which provides an easy access to the technologies described in this paper.

It is important to note, that the proposed framework is fully based on components available through open source licenses (GPL). Additionally, all Web Services listed in the framework are compliant to open standards. This openness of our approach is an essential aspect, since components can be replaced and extended for future requirements. We want to stress that the openness results finally in a sustainable approach for building SOAs and prevents from vendor lock-in (license and interface-wise). The presented framework is published as open source by the Geoprocessing Community at the 52°North Open Source initiative. All the implementations are part of the 52°North WPS framework (in the following referred to as 52°North

WPS).[3] A thorough performance evaluation of the framework is outside the scope of this paper, but e.g. Scholten et al. (2006) review the performance of distributed geoprocessing applications.

At first, the paper provides a review of the basic concepts applied in the framework: OGC WPS interface specification, Grid Computing and geoprocessing workflows. After that the paper presents the combined approach of Grid Computing and geoprocessing workflows. To build sustainable workflows, existing geoprocessing functionality incorporated in stand-alone desktop GIS has to be exposed by Web Services. In this context the wrapping of GIS functionality with a WPS interface is discussed based on the experiences with GRASS GIS. In Sect. 1.4 the proposed framework is applied to a real-world scenario for proving the introduced concepts. The paper ends with a discussion and a conclusion about the framework.

1.2 Technical Background and Related Work

This section provides a review of related work in the context of OGC Web Processing Service, Grid Computing and Web Service Orchestration.

1.2.1 OGC Web Processing Service

The OGC Web Processing Service interface specification describes a standardized method to publish and execute web-based processes for any type of geoprocess. According to the WPS interface specification, a process is defined as any calculation operating on spatially referenced data.

In detail, the WPS interface specification describes three operations, which are all handled in a stateless manner: *GetCapabilities*, *DescribeProcess* and *Execute*. *GetCapabilities* is common to any type of OGC Web Service and returns service metadata. In case of WPS, it also returns a brief description of the processes offered by the specific service instance. To get more information about the hosted processes, the WPS provides process metadata through the *DescribeProcess* operation. This operation describes all parameters, which are required to run the process. Based on this information the client can perform the *Execute* operation upon the designated process. As every OGC Web Service, the WPS communicates

[3] The Geoprocessing community at 52°North: http://52north.org/wps

through HTTP-GET and HTTP-POST based on an OGC-specific XML-message encoding.

Besides this basic communication pattern, the WPS interface provides functionality for scalable processing such as asynchronous processing (implemented using the pull model), storing of process results and processing of data references encoded as URLs. The application of URL references as input for specific processes is a promising feature, as it limits the volume of data sent between client and service and allows the service to apply specific caching strategies. The service retrieves the data once and reuses it multiple times, by using the reference as an identifier for data.

1.2.2 Grid Computing

The term *Grid Computing* is a diffuse phrase and there are many definitions available. This lack of a single and precise definition leads to a broad variety of understandings of this term. However, the term Grid Computing is frequently used when linking computational resources to solve time- and resource-intensive computations. Following the argumentation of Foster and Kesselmann (1998), the term Grid Computing must be evaluated in terms of the applications, business value, and scientific results that it delivers, not its architecture. Moreover, Foster characterizes in his famous three point checklist[4] a system as a Grid that:

1. Coordinates resources that are not subject to centralized control
2. Is using standard, open, general-purpose protocols and interface
3. Delivers nontrivial qualities of service

The emerging term *Cloud Computing* overlaps with some concepts of distributed and Grid Computing (Hartig 2008). It uses a cloud metaphor to represent the internet or other large networking infrastructures. The paradigm behind the buzzword foresees a future in which data and computations will be handled by distributed facilities operated by third-party resources (Foster et al. 2008). The key characteristics of the cloud are the ability to scale and to provide computing power on-demand in a cost efficient way and the ability of the consumer to benefit from these resources without having to manage the underlying complexity of the technology. These characteristics lead to a set of core value propositions like scalability on demand, streamlining the data center, improving

---

[4] Ian Foster, "What is the Grid? A three point checklist": http://www.gridtoday.com/02/0722/100136.html

business processes and minimizing startup costs for new business model.[5] In essence, Cloud Computing is not a completely new concept. Moreover, it combines a family of well known and established methods and technologies (for example SaaS as a model for software packaging and deployment, Virtualization as an efficient hosting platform and Grid Computing as a backbone for providing sufficient computational power). Besides, it describes a paradigm of outsourcing applications and specific tasks to a scalable infrastructure and therefore consequently enabling new business models with less up-front investments.

In the context of linking geospatial applications and SDIs with Grid Computing only little research has been conducted. Baranski (2008) and Lanig et al. (2008) extended the work of Di et al. (2003) and accomplished first experiments using Grid Computing technology for improving processing performance by distributing and parallel execution of processes. In the OWS-6 testbed (OWS-6 2009) the adoption of Open Grid Forum (OGF) standards for the WPS interface specification was reviewed. Recently, the GDI-GRID[6] project has started focusing on the integration and processing of geodata and services in a grid infrastructure. In particular, it addresses the implementation of security mechanisms and definition of service interfaces. The purpose of SEE-GEO[7] project is to provide access to the EDINA[8] national data enter hosting geospatial services running on the UK National Grid Service.[9] The SLA4D-GRID[10] project focuses on the development of a common software stack for managing and enforcing Service Level Agreements (SLA) in the German grid infrastructure D-GRID via advanced resource reservation mechanisms. A SLA-enabled WPS is one important use case in this research project.

1.2.3 Web Service Orchestration and Workflows

Complex business processes often require the functionality of multiple tasks. These tasks could be often realized as Web Services which can be composed to one workflow to represent the desired business process. Alonso et al. (2003) speak of a composite service and entitle the act of

[5] The Open Cloud Manifesto homepage: http://www.opencloudmanifesto.org

[6] GDI-Grid project homepage: http://www.gdi-grid.de/

[7] SEE and SAW-GEO project homepage: http://edina.ac.uk/projects/seesaw/index.html

[8] JISC national academic data centre based at the University of Edinburgh: http://edina.ac.uk/

[9] UK National Grid Service (NGS): http://www.grid-support.ac.uk/

[10] SLA4D-GRID project homepage: http://www.sla4d-grid.de

combining Web Services as *Web Service Composition*. The interaction of the Web Service in the workflow can be handled by *Web Service Orchestration*, which is a description of how combined Web Services interact on the message level (Pelz 2003). This description includes the business-logic and execution order. Thereby it can be distinguished from *Web Service Choreography*, which has a more collaborative character and is more focused on the public message exchange between multiple parties. Since *Web Service Choreography* specifies an explicit message interaction protocol, each service knows its predecessor and successor.

This differs from *Web Service Orchestration*, which is often realized by a central orchestration engine coordinating the Web Service interaction according to a predefined workflow description. Thereby, this approach allows the Web Services to be held loosely coupled (Weerawarana et al. 2006).

In this respect, the Business Process Execution Language (BPEL) has advanced to a de-facto standard in the mainstream IT-world. BPEL is an OASIS standard[11] for describing workflows based on elementary tasks implemented as Web Services using an XML-encoding. In particular, these so-called BPEL scripts describe the roles involved in the message exchange, supported port types and orchestration information of a process. Additionally, specific binding and deployment issues remain at the Web Service and not at the BPEL process itself. All these aspects make BPEL an appropriate candidate to represent geoprocessing workflows in SOAs.

1.3 An Approach Towards Real-Time and Distributed Geoprocessing

This section presents the developed approach towards real-time and distributed geoprocessing. At first, the wrapping of stand-alone geoprocessing functionality as a Web Service is explained. Secondly, an approach for the gridification OGC Web Services is presented. Finally, both concepts are combined into an approach for geoprocessing workflows.

[11] Web Services Business Process Execution Language, Version 2.0: http://docs.oasis-open.org/wsbpel/2.0/OS/wsbpel-v2.0-OS.html

1.3.1 GIS Wrapping

A wrapper is responsible for transforming incoming service requests into adequate calls for the underlying (legacy) system without any modifications of the wrapping interface (i.e. WPS interface) (Erl 2005; Papazoglou and van den Heuvel 2007). In this section, requirements to wrap geoprocessing functionality are described. Additionally, an architecture for such a wrapper will be presented based on the popular stand-alone open source GRASS GIS and the 52°North WPS framework.

Table 1.1 provides an overview of requirements for successfully wrapping geoprocessing functionality. To wrap a large number of existing functionality, automation is a key aspect. In particular, the automated extraction of the description of the functionality and the automated forward of the service call is a key requirement. This is achieved through a programmable interface (*shell and Python scripting* in GRASS GIS), which is accessible for the wrapping framework. All other requirements are optional.

Table 1.1. Mandatory and optional characteristics for successful wrapping GIS functionality

Requirement	Type of requirement
Programmable interface for unsupervised processing	Mandatory
Multitasking	Optional
Machine-readable interface descriptions	Optional (for automatic creation of Describe Process documents mandatory)
Determination of coordinate reference systems	Optional (technical level) / mandatory (regarding correct results)

Workspaces are common for importing, processing data and finally storing the results in GIS such as for instance in GRASS. However, this concept does not match the stateless fashion of WPS. Therefore, a temporary workspace has to be created. This implies that the data are not usable to the process afterwards and should be deleted after sending the response. This leads to the architecture shown in Fig. 1.1.

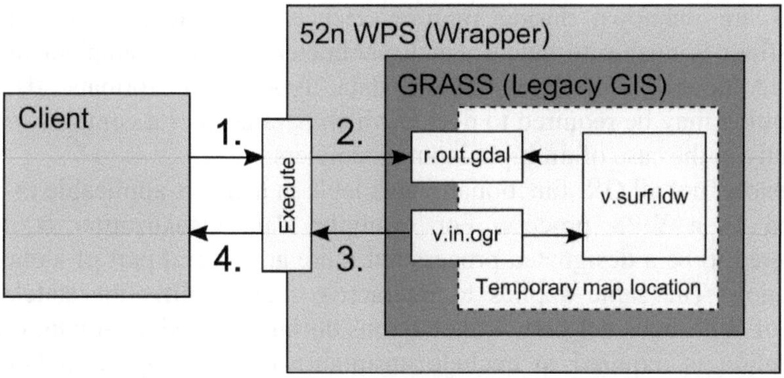

Fig. 1.1. Architecture for wrapping GRASS with a WPS interface. An Execute request for an Inverse Distance Weighting (IDW) interpolation functionality of GRASS is sent and processed

The architecture consists of three components: the client, the 52°North WPS (wrapper) and the GRASS GIS (legacy GIS). An Execute request is received from the client (Fig. 1.1, Step 1). In this example Inverse Distance Weighting (IDW) functionality is requested, which will also be used in the scenario (Sect. 1.4). The wrapper examines the request and creates an adequate shell script to invoke the underlying GRASS GIS (Step 2).This includes pre-configuring of GRASS, importing the input data, invoking the desired functionality and exporting the result. A reference to a temporary file is returned back to the wrapper (Step 3), and handed back to the client (Step 4). With minor modifications this architecture is applicable for other GIS as well. The type of programming interface might change from shell scripts to for example a programmable Python interface, but the architecture remains the same.

If the GIS provides machine-readable interface descriptions for its functionality, it is possible to transform these descriptions into Describe Process documents automatically. This is especially convenient if a full set of GIS functionality (in GRASS there are about 400 different pieces of functionality) should be offered by a WPS. To transform the XML-based descriptions of GRASS the eXtensible Stylesheet Language (XSL) is used. An XSL-transformation (XSLT) filter creates the DescribeProcess document elements via if-then-else statements, case differentiation and other information (e.g. MIME type for complex parameters).The XSLT-filter has to be implemented only once and can be applied to extract any process description fully-automated in the future.

Regarding GRASS, there are two semantic problems which constrain this fully-automated approach. Firstly, GRASS uses a non-typed parameter input system. This has certain drawbacks for WPS, as datasets and their formats are unknown during runtime. Hence, the machine controlled XSLT-filter is unable to detect which parameter has to be supplemented with additional information about data types and formats. User intervention may be required to decide which parameter is a complex one, especially in the case of multiple input parameters.

Secondly, not all GIS functionality available in a GIS is applicable to be exposed as a WPS process. For example, data visualization is not considered to be a designated process but more an integral part of a client application. The same applies to interactive functionality: the stateless nature of WPS does not permit interactions during processing. Again, user intervention is required to exclude unsuitable functionality from being offered through the WPS.

Nevertheless, this semi-automatic approach to create process descriptions is much more convenient than creating a full set of DescribeProcess documents manually.

Further details about the architecture and implementation can be found in Brauner and Schaeffer (2008).

1.3.2 Grid Computing

Two potential ways can be identified to adapt OGC Web Services (OWS) to domain-specific requirements of a Grid Computing environment and to make use of Open Grid Forum (OGF) and related specifications, concepts and their implementations.

At first, the OWS interacts with distributed computing resource in the backend (encapsulating other resources). In this simple approach the OWS could use other grid services, for example, to send calculation jobs to compute resources or for accessing distributed databases based on well defined mechanisms. In the course of this work this approach will be referred to as “encapsulation” as the clients of the OWS are not exposed to any changes of the OWS interface. The encapsulation fosters the performance of distributed time- and resource-intensive processes and the access of large and distributed datasets.

Secondly, the OWS is integrated within other Grid Computing environments. Therefore, the service has to be fully embedded – mostly as a stateful web service – into the grid middleware. This approach offers opportunities to realize benefit from existing middleware solutions such as intelligent workflow management, reliability, security and other QoS. In the course of this work this approach is referred to as “integration” as the OWS becomes a service that clients may utilize alongside others in the Grid Computing environment. Current OGC standards define typically a stateless message exchange via plain HTTP-GET and HTTP-POST protocol, but most of common grid middleware applications are based on the stateful Web Service Resource Framework (WSRF). Therefore, regarding the integration, the gridified OWS must be accessed via an OGC-compliant proxy component.

The presented implementation of an encapsulating WPS (in the course of this work referred as WPS-G) is based on the 52°North WPS and the open source UNICORE grid middleware.[12] A brief architecture overview is shown in Fig. 1.2.

For executing distributable processes, the implementation divides the processing problem into smaller sub-problems by splitting up the input data into smaller chunks. Then these chunks and necessary application binaries will be distributed upon several computational nodes in the grid. The number of computation nodes depends on the complexity of the process and can be either adjusted manually or assessed automatically. After preparing each computational node, the application binaries will be executed in parallel. During this execution phase, the CPU at the WPS server side is idle and waiting for process results or other incoming requests. When all computational nodes have performed the execution successfully, the WPS fetches all result datasets and merges them together.

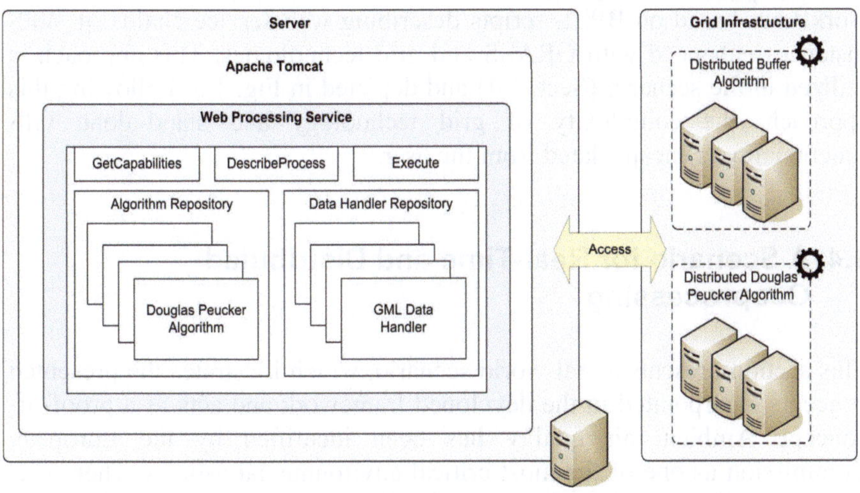

Fig. 1.2. An architectural overview of the gridified WPS implementation

[12] UNICORE grid middleware: http://www.unicore.eu

1.3.3 Scalable Geoprocessing Workflows

This section describes the integration of the GRASS-based (Sect. 1.3.1) and gridified (Sect. 1.3.2) WPS into a geoprocessing workflow. Since the interfaces of the gridified and GRASS-based WPS instances were not changed. Hence, the WPS-G and the GRASS-based WPS can be exactly treated like any OGC compliant WPS. Furthermore, a WPS is a Web Service, which can be used to create geoprocessing workflows in SOAs (Sect. 1.2). For instance, Weiser and Zipf (2007) showed this for simple OGC Web Services without a desktop GIS backend or grid technology. The Orchestra project developed similar approaches for the orchestration of geospatial web services, but neither incorporated grid technologies nor used standard OGC web services as frontend to existing geoprocessing functionality (Usländer 2007). Besides, a separate class of geospatial web services was developed for the Orchestra project, which has not been adopted by international standardization bodies yet.

Our approach goes beyond that and allows the reuse of existing geoprocessing workflows by exposing the geoprocessing workflow as a simple WPS process using the WPS-T approach (T stands for transactional) (Schaeffer 2008). This approach enables the dynamic deployment of arbitrary geoprocessing workflows. In our work the approach applies workflows based on BPEL scripts describing web service chains of WPS instances enhanced with GRASS and grid technologies. This approach is utilized in the scenario (Sect. 1.4) and depicted in Fig. 1.3. Following this approach, the complexity of grid technology and stand-alone GIS functionality is encapsulated from the user.

1.4 A Scenario for Real-Time and Distributed Geoprocessing

This section presents a real-world scenario, which illustrates the presented concepts incorporated in the developed framework and acts as a proof-of-concept. Ambient air quality has been identified by the European Commission as one of the most critical environmental aspects. Therefore, it is one of the major objectives in the Sixth Environmental Action Programme.[13] Through legislation, the European Union has also been active in this field for several decades. With the Council Directives

[13] The Sixth Environment Action Programme of the European Community: http://ec.europa.eu/environment/newprg/index.htm

1999/30/EC and 1996/62/EC strict air quality limits for several air quality parameters were specified, which have to be adapted by national law by all member states.

This scenario addresses Particulate Matter (PM10) as one of the most hazardous aerosols. Several studies have researched the correlation of PM10 and health effects. For instance, Pope et al. (2002) found that for each rise of 10 $\mu g/m^3$ the lung cancer rate increases by 4–8%. Hence, the real-time monitoring of PM10 is a crucial task to support decision makers in protecting public health and to comply with the EU regulations. In order to automate this business process, a model realized as a geoprocessing workflow will be developed to monitor real-time spatial distributions of PM10 by applying the technical concepts introduced in this paper.

Figure 1.3 presents the deployed workflow. It was modelled with the 52°North Workflow Modeler, which automatically creates WSDL documents for the workflow partners and describes the workflow via BPEL. While for instance Weiser and Zipf (2007) also showed the orchestration of OGC Web Services with BPEL, they had to build the WSDL manually. Our approach goes beyond that and exposes the deployed workflow as a simple WPS process using the WPS-T approach (Sect. 1.3.3). The client on the left-hand side (Fig. 1.3) is now capable to invoke the workflow like any other WPS process (Step 1). Only the reference to the sensor data has to be included in the request. In our case, the 52°North WPS uDig client application (Schaeffer and Foerster 2008) can be used to invoke the service and visualize the results (Fig. 1.4).

After invoking the workflow represented by a WPS process, hosted on the WPS-T, the underlying workflow engine is triggered in the backend (Step 2). Upon then, the workflow engine orchestrates the workflow and invokes each of the workflow partners. At first, a WPS filter process, which takes a URL from an OGC Sensor Observation Service (SOS) as input (Step 3) is triggered. The SOS instance is located on top of a sensor network and allows the client to query the observed phenomena in a standardized format. In this scenario real-time air quality data from the state of North-Rhine Westphalia, located in the western part of Germany, is available through the SOS instance. The SOS instance provides up-to-date data of 56 monitoring stations distributed through out this region. Internally, the WPS queries the SOS for its latest available sensor data regarding this specific region (Step 4). Next, the retrieved sensor data is converted into the common Geography Markup Language (GML) format (Step 5), which can be used for further processing. In particular, the geometry and observed values are extracted and new GML point features are created. This process is performed by a gridified WPS, which delegates the conversion task to its underlying grid middleware due to high computational complexity. The following step (Step 6) takes the resulting point data from the second process as input and interpolates them with the GRASS Inverse Distance Weighting (IDW) method to a resulting GeoTIFF raster. In the last step (Step 7) the GeoTIFF raster becomes visualized according to the GRASS color scheme for air pollution is applied, in which the color red represents values above a specific threshold (40 μg PM10 /m³ air as the current EU legislation limit). The output is visualized as a new layer in the open source uDig client (Fig. 1.4).

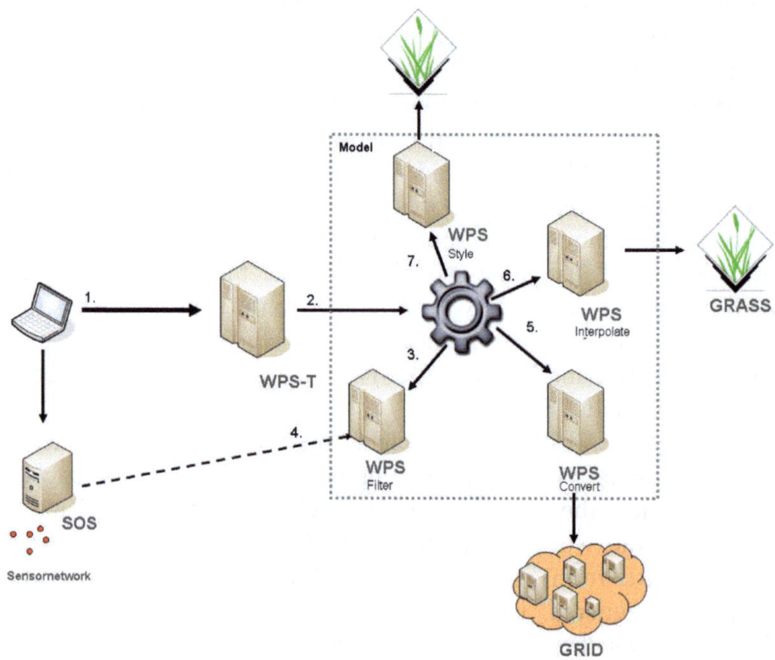

Fig. 1.3. Deployed workflow for the described scenario

As the workflow analyses latest data in (almost) real-time and can be triggered on-demand, it allows the decision makers to monitor the current air quality and thereby to take actions accordingly based on the latest information in real-time. Additionally, it is possible to compare the phenomena over time and thereby get analyze its behaviour more thoroughly.

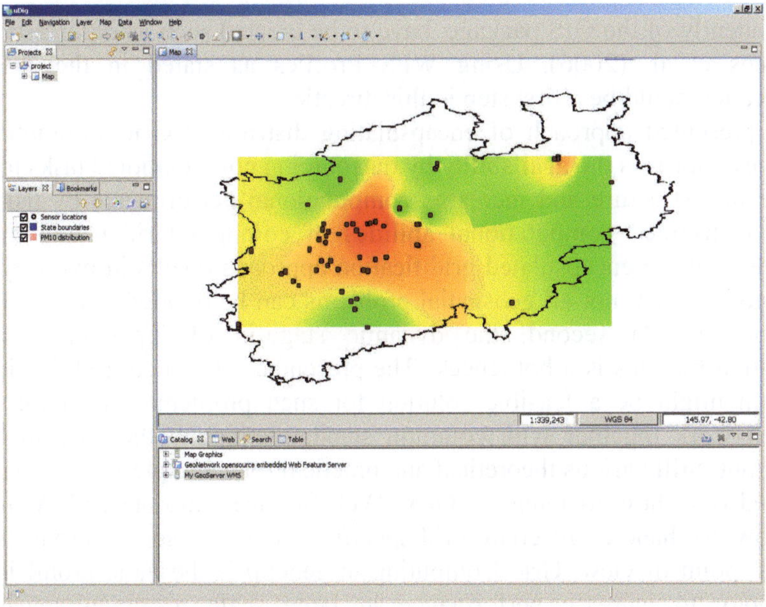

Fig. 1.4. Workflow results visualized with states boundary of North-Rhine Westphalia and the sensor location. The red colored areas imply a high amount of PM10 (40 μg PM10 /m³ air), which violates current EU regulations

1.5 Discussion and Conclusion

This paper describes an open source framework for building geoprocessing workflows using real-time geodata based on distributed Web Services. First, existing geoprocessing functionality of stand-alone GIS is exposed as Web Services. Second, computational intense processes are delegated to a Grid Computing environment. Since both approaches do not change the standardized WPS interface, they can be easily applied in a geoprocessing workflow. Moreover, the framework includes a mechanism to expose geoprocessing workflows as single processes through WPS interface (WPS-T). All the mechanisms of the framework are utilized to realize a real-time scenario for air quality monitoring (Sect. 1.4).

Besides the general geoprocessing research topics elaborated in Brauner et al. (2009), this paper shows that wrapping geoprocessing functionality with a WPS interface is a promising concept to reuse existing and reliable functionality in a Web Service context. However the semantic gap between the user and the process description still remains. In particular, taxonomies for geoprocesses to classify and identify functionality and its parameters, independently of the GIS package, have to be created such as described in Lemmens et al. (2006). Using WPS Profiles as stated in the WPS specification could be a first step in this direction.

The presented approach of encapsulating distributed Grid Computing resources improves the availability by outsourcing computational tasks to a failsafe infrastructure and increases computational performance by using shared distributed computational abilities of common Grid Computing methods. But the encapsulated gridification approach results in two major problems. At first, not all geospatial problems can be divided into smaller sub-problems. At second, the dynamic stage-in of input data and application binaries is a bottleneck. The presented integrated gridification approach might be a feasible solution for such problems, but a lot of research work together with common standardization bodies has to be carried out. Still various theoretical and practical relevant questions remain unsolved (e.g. how to integrate OGC Web Services into stateful WSRF and how to handle different grid-specific security issues). From the authors' point of view, Grid Computing in general is the most promising technology to increase performance in large scale resource-intensive geoprocessing tasks.

The deployment of workflows using real-time data from distributed services allows the service provider to model such workflows only once but execute them anytime by a client application. This is highly relevant for complex workflows and significant for real-time risk management applications as demonstrated by the scenario. Additionally, the workflow can be integrated as another building block into a higher level workflow since it is exposed as a single WPS process (WPS-T). Finally, this results in nested workflows, which are again based on workflows. Even though in our case only WPS instances are incorporated, the presented approach for establishing workflows can be applied to any type of Web Service interface because the approach is based on the mainstream IT standard BPEL. Asynchronous processing and the use of references have to be taken into account to improve the performance of the designed workflows in the future.

Reference

Alameh N (2003) Chaining geographic information web services. IEEE Internet Comput 07:22–29. doi:10.1109/MIC.2003.1232514

Alonso G, Casati F, Kuno H (2003) Web services. Concepts, architectures and application. Springer, Berlin

Baranski B (2008) Grid computing enabled web processing service. GI-Days 2008, University of Münster, Germany

Brauner J, Schaeffer B (2008) Integration of GRASS functionality in web based SDI service chains. Academic paper at FOOSS4G 2009, Cape Town, South Africa

Brauner J, Foerster T, Schaeffer B, Baranski B (2009) Towards a research agenda for geoprocessing services. In: 12th AGILE international conference on geographic information science, Hannover, Germany

Di L, Chen A, Yang W, Zhao P (2003) The integration of grid technology with OGC web services (OWS) in NWGISS for NASA EOS Data. GGF8 & HPDC12. 24–27 June, Seattle

Erl T (2005) Service-oriented architecture: concepts, technology, and design. Prentice Hall PTR, Upper Saddle River

Foster I, Kesselman C (1998) The grid: a blueprint for a new computing infrastructure. Morgan Kaufmann, San Francisco

Foster I, Zhao Y, Raicu I, Lu S (2008) Cloud computing and grid computing 360-degree compared. http://arxiv.org/abs/0901.0131. Accessed 26 May 2009

Hartig K (2008) What is cloud computing? The cloud is a virtualization of resources that maintains and manages itself. In: .NET Developers Journal, SYS500 CON Media

Kiehle C, Greve K, Heier C (2007) Requirements for next generation spatial data infrastructures-standardized web based geoprocessing and web service orchestration. Trans GIS 11(6):819–834. doi:10.1111/j.1467-9671.2007.01076.x

Lanig S, Schilling A, Stollberg B, Zipf A (2008) Towards standards-based processing of digital elevation models for grid computing through web processing service (WPS). The 2008 international conference on computational science and its applications (ICCSA2008), Perugia, doi:10.1007/978-3-540-69848-7_17

Lemmens R, Wytzisk A, de By R, Granell C, Gould M, van Oosterom P (2006) Integrating semantic and syntactic descriptions to chain geographic services. IEEE Internet Comput 10:42–52. doi:10.1109/MIC.2006.106

Masser I (2005) GIS worlds: creating spatial data infrastructures, 1st edn. ESRI Press, Redlands

OWS-6 (2009) OWS-6 WPS grid processing profile engineering report. OGC 09-041 (Forthcoming)

Papazoglou MP, van den Heuvel W-J (2007) Service oriented architectures: approaches, technologies and research issues. VLDB J 16(3):389–415. doi:10.1007/s00778-007-0044-3

Pelz C (2003) Web services orchestration and choreography. IEEE Comput 36(8):46–52. doi:10.1109/MC.2003.1236471

Pope C, Burnett R, Thun M, Calle E, Krewski D (2002) Lung cancer, cardiopulmonary mortality, and long-term exposure to fine particulate air pollution. Med Assoc 287:132–1141. doi:10.1001/jama.287.9.1132

Schaeffer B (2008) Towards a transactional web processing service (WPS-T). In: Pebesma E, Bishr M, Bartoschek T (eds) Proc 6th GI Days June 16–18, 2008. Institute for Geoinformatics, Muenster, pp 91–117

Schaeffer B, Foerster T (2008) A client for distributed geo-processing and workflow design. J Location Based Services 2(3):194–210. doi:doi:10.1080/17489720802558491

Scholten M, Klamma R, Kiehle C (2006) Evaluating performance in spatial data infrastructures for geoprocessing. IEEE Internet Comput 10(5):34–41. doi:doi:10.1109/MIC.2006.97

Usländer T (ed) (2007) Reference model for the ORCHESTRA architecture. http://www.eu-orchestra.org/download.php?file=docs/RM-OA/RM-OA-V2-Rev-2.0.pdf. Accessed 26 May 2009

van der Aalst W, van Hee K (2004) Workflow management: models, methods and systems. MIT Press, Cambridge

Weerawarana S, Curbera F, Leymann F, Storey T, Ferguson D (2006) Web services platform architecture: SOAP, WSDL, WS-policy, WS-addressing, WS-BPEL, WS-reliable messaging and more, 4th edn. Prentice Hall/PTR, Upper Saddle River

Weiser A, Zipf A (2007) Web service orchestration (WSO) of OGC Web Services (OWS) for Disaster Management. In: Li J, Zlatanova S, Fabbri A (eds) Lecture notes in geoinformation and cartography. Springer, New York, pp 239–254

Chapter 2 GeOxygene: Built on Top of the Expertise of the French NMA to Host and Share Advanced GI Science Research Results

Bénédicte Bucher, Mickaël Brasebin, Elodie Buard, Eric Grosso, Sébastien Mustière, Julien Perret

Institut Géographique National, Laboratoire COGIT 73, avenue de Paris 94160 Saint Mandé, France
{benedicte.bucher, mickael.brasebin, elodie.buard, eric.grosso, sebastien.mustiere, julien.perret}@ign.fr

Abstract. Back to the 80s and 90s, researchers at the COGIT laboratory of the French National Mapping Agency often developed their own vector platforms dedicated to their respective research purposes like generalization, integration or 3D. This multiplicity of platforms became an obstacle to many important research activities like capitalizing and sharing. In 2000, based on ISO/OGC standard for data exchange, the COGIT laboratory decided to invest in the development of an ad hoc interoperable GIS platform that would host most of its research developments. This platform, called GeOxygene, was designed to have a rich data model and a genuine programming language to manipulate data. It facilitates the design and prototyping of processes which geographical data model are object oriented. It is also intended to facilitate the sharing of knowledge and programming efforts with other research labs. It is also shared through regular open source releases. This paper highlights the relevance for a research team like COGIT to contribute to Open Source software.

2.1 Introduction

The research department of the French National Mapping Agency, IGN, is composed of four laboratories. One is working on geodesy, another on hardware issues (like digital cameras), another on image processing and the

E. Bocher and M. Neteler (eds.), *Geospatial Free and Open Source Software in the 21st Century*, Lecture Notes in Geoinformation and Cartography,
DOI 10.1007/978-3-642-10595-1_2, © Springer-Verlag Berlin Heidelberg 2012

last one, COGIT,[1] on the management and usage of interpreted geodata. Some COGIT research activities address IGN production which include automating generalization and label placements, or automated data matching and integration for quality controls or update process. Other COGIT research activities target external end users' needs like providing better interfaces for users to access relevant data and processes or providing better data models and methods for specific application domains.

In the 80s and 90s, using existing proprietary GIS software dedicated to vector data was an issue for COGIT research activities because these software, though they provide 'toolboxes' and APIs, were not completely open, i.e. their internal data models were not entirely visible. Furthermore, these software were often associated with *ad hoc* programming languages (e.g. Avenue language for ESRI ArcView), leading to additional training of developers and poor code re-usability. In other words, we lacked in 2000 an Open Source software to handle vector geographic data associated with an object oriented programming language (C++ or Java).

This is the reason why, at that time, COGIT researchers often developed their own platforms dedicated to their respective research purposes like generalization, integration or 3D. Depending on their research area they would use different programming languages like ADA, C, LISP and so on. This multiplicity of platforms became an obstacle to many important research activities like pooling and sharing geoknowledge and geo-applications. For example, as detailed in further sections, four PhD theses have been lead since 1997 on 3D modelling using different platforms. Besides, these were non-standard proprietary, independent or non interoperable platforms which did not encourage exchanges with other researchers.

In 2000, the COGIT laboratory decided to invest in the development of an *ad hoc* GIS platform that would host research developments, except for those related to generalization which were hosted on a dedicated platform designed in the context of a European project. This *ad hoc* GIS platform was designed to have a rich data model and a genuine programming language to manipulate data because COGIT expertise lies there: the modelling of geographical data and knowledge to improve and automate specific processes. This platform is called GeOxygene (Badard and Braun 2003) and is released under the terms of the LGPL[2] (GNU Lesser General Public License) license. It uses standard ISO/OGC models to represent geographic information. It also uses the wide spread Java language which

---

[1] Acronym for Conception Objet et Généralisation de l'Information Topographique (http://recherche.ign.fr/labos/cogit/).

[2] http://www.gnu.org/licenses/lgpl.html

offers interfaces to native languages. There are regular Open Source releases of GeOxygene which contain the core packages and some specific packages (basic functions, data matching, 3D, semiology, etc.) that IGN has decided to share as Open Source software, like for instance the data matching package.

The remainder of the paper is not limited to the current Open Source release of GeOxygene. We rather describe the usage of GeOxygene in our laboratory and then focus on different specific packages. Some of these packages are part of GeOxygene Open Source release.

2.2 GeOxygene: A Platform to Host and Share Research Developments

GeOxygene is a development platform and not a generalist GIS platform. It is dedicated to hosting programs developed by COGIT researchers. Consequently, not much attention was initially paid to the development of GIS support functions which were not targeted by these researches, like import, export and visualization functions. Rather, the data model has been the main focus. GeOxygene's specificity is to load geographical data into an object oriented model and to seamlessly manage the binding between a storage model and a working object model. In other words, GeOxygene facilitates the design and prototyping of processes which geographical data models are object oriented.

An important aspect of GeOxygene is the sharing and capitalizing of knowledge. Several strategies exist in order to share knowledge based on GeOxygene.

The first strategy consists in using a revision control system so that all researchers may load and see the up-to-date code of every researcher in the laboratory. To enhance code sharing based on such tools, catalogs are being developed to index all these libraries by the functions and models they support (Abd el Kader and Bucher 2006).

Another strategy is the dissemination of an open source version of GeOxygene among the community of researchers. Typically, this is a good incentive to keep effective collaboration with former COGIT researchers. COGIT's GeOxygene version is not fully synchronized with the Open Source version (hosted on SourceForge[3]). Not fully synchronized means that relevant modules are released as they reach a certain level of maturity and usability. Besides, this open sourcing allows the definition of new

[3] http://oxygene-project.sourceforge.net/

collaborations based on GeOxygene like the GeOpenSim project funded by the French National Research Agency (ANR). GeOpenSim aims at designing an Open Source platform for the simulation of urban dynamics (Perret et al. 2009).

The last strategy is the development of interoperable Web applications. It has been targeted from the early beginning of GeOxygene's development (Badard and Braun 2004). Indeed, using ISO/OGC data models to structure geographical data ensures that input and output data can be serialized into standard XML schemas. Developing geographical Web Services is an important activity for IGN for the upcoming information infrastructures. Balley (2004) addressed a first challenge in developing interoperable services above structured data: assisted schema transformation. Other works focus on the development of cartographic Web services to enhance the quality of maps (Bucher et al. 2007). Last, some work aims at developing Web services for integrating historical data, like old maps, into recent georeferenced frameworks (Grosso et al. 2009).

The next sections depict specific GeOxygene libraries.

2.3 GeOxygene Libraries for Data Matching

Due to the increasing number and diversity of available geographic information, there is a growing need for the integration of different databases into a single one making explicit the relationships between them. This requires data matching tools that explicit the links between homologous objects, i.e. objects from the different databases that correspond to the same real world entity. Matching homologous objects belonging to different databases is useful in many contexts (Kilpeläinen, 2000; Hampe and Sester 2002; Mustière and van Smaalen 2007): it facilitates propagation of updates between databases; it paves the way to the assessment of the quality of one database compared to another one taken as a reference; it allows users to make multi-criteria analyses combining information from various sources; it helps the development of efficient data visualization tools navigating between points of views.

In this context, the COGIT laboratory has developed data matching tools specifically dedicated to matching networks with different levels of detail (Mustière and Devogele 2008). These tools are based on the comparison of geometrical and topological aspects of networks. They are in particular able to match one node of the less detailed network to a set of nodes and edges representing a complex crossroad in the other networks. Figure 2.1 illustrates different cases where a node of the less detailed network is matched, respectively, to a roundabout (Fig. 2.1, 1), a shifted junction (Fig. 2.1, 2) or a square junction (Fig. 2.1, 3).

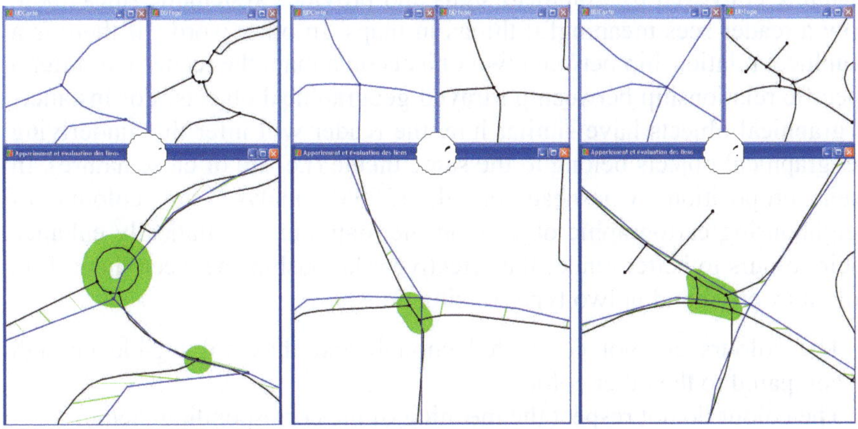

Fig. 2.1. Data matching results. In each of the three subfigures, the less detailed network (*top left*), the most detailed one (*top right*), and their superimposition and matching (*bottom*) are represented

This tool has been released as Open Source in 2008. This concerns the data matching algorithms themselves, but also OpenJump plugins for the simultaneous visualisation of different networks, as shown in Fig. 2.1. Compared to existing tools like OpenJump Java Conflation Suite, GeOxygene algorithms focus on the analysis of topological relationships in networks, and are especially dedicated to the matching of networks at different scales, leading to numerous one-to-many matching links. It has already been tested by several institutions such as the Belgium National Geographic Institute and the Ordnance Survey.

Other works on this subject concern a new approach for data matching based on a multi-criteria analysis of feature characteristics, including descriptive attributes. It uses techniques inspired by the so-called evidence theory to combine these criteria (Olteanu and Mustière 2008).

2.4 Management of Colour Contrasts in Maps

Another domain where IGN and COGIT have a solid expertise is the drawing of legible maps based on cartographic theory, including semiology of graphics. Indeed, part of IGN's vocation is the design and production of maps. Drawing legible maps is not only an issue of accessing and drawing data. Chesneau et al. (2005) aimed at providing automatic mechanisms to enhance the legibility of risk maps based on the semiology of graphics theory (Bertin 1983). This theory is a framework for the analysis of the meaning conveyed by graphical signs and provides systematic rules about how a reader sees meaningful things in maps. In other words, if there is a graphical relationship between two objects on a map, the reader will infer a specific relationship between portrayed geographical objects. For instance, if graphical objects have similar hue, the reader will infer that underlying geographical objects belong to the same theme (i.e. are of close nature). In their proposition, Chesneau et al. (2005) analyse the colours of neighbouring cartographic objects on the map and automatically enhance their colours to better render the effective relationships between them. Bad colours were found in two types of situations:

- The colours are not contrasted enough and thus not legible enough compared to the other colours.
- The colour do not respect the meaning of the corresponding theme.

The relative positions of all used colours give meaning to the themes, according to their differences. These relationships set up a comprehensive understanding of the whole map.

To facilitate colour analysis, we integrate in the process a colour space reduction obtained by creating a chromatic circle of 12 hues, each derived into 7 values. This repartition is used to visualise relationships: association and difference are represented respectively by low angle and high angle between hues, whereas order is represented by value increasing within the same hue. Following works focus on the determination of contrasts marks for each couple of colours in this chromatic circle (Buard and Ruas 2007). These marks have been obtained through user tests where cartographic experts were asked to give scores to specific colour contrasts: hue and value. These scores are further used to compute contrast marks of hue and value between two neighbouring objects on the map.

The module graphical user interface is designed as an OpenJump plugin. It can process any data with any colour. The module enables the user to test his own data and to be creative. It has been successfully tested on risk data which are quite complex to portray.

To activate the analysis and improvement of colour contrasts, we use two menus. The analysis menu is activated only once: it deals with identifying neighbouring objects on the map. The improvement menu can be activated as many times as wanted, knowing that each time a new state with a change of colour appears. In this way, the state cycle is not automated but supervised by the user. He can stop when he is satisfied.

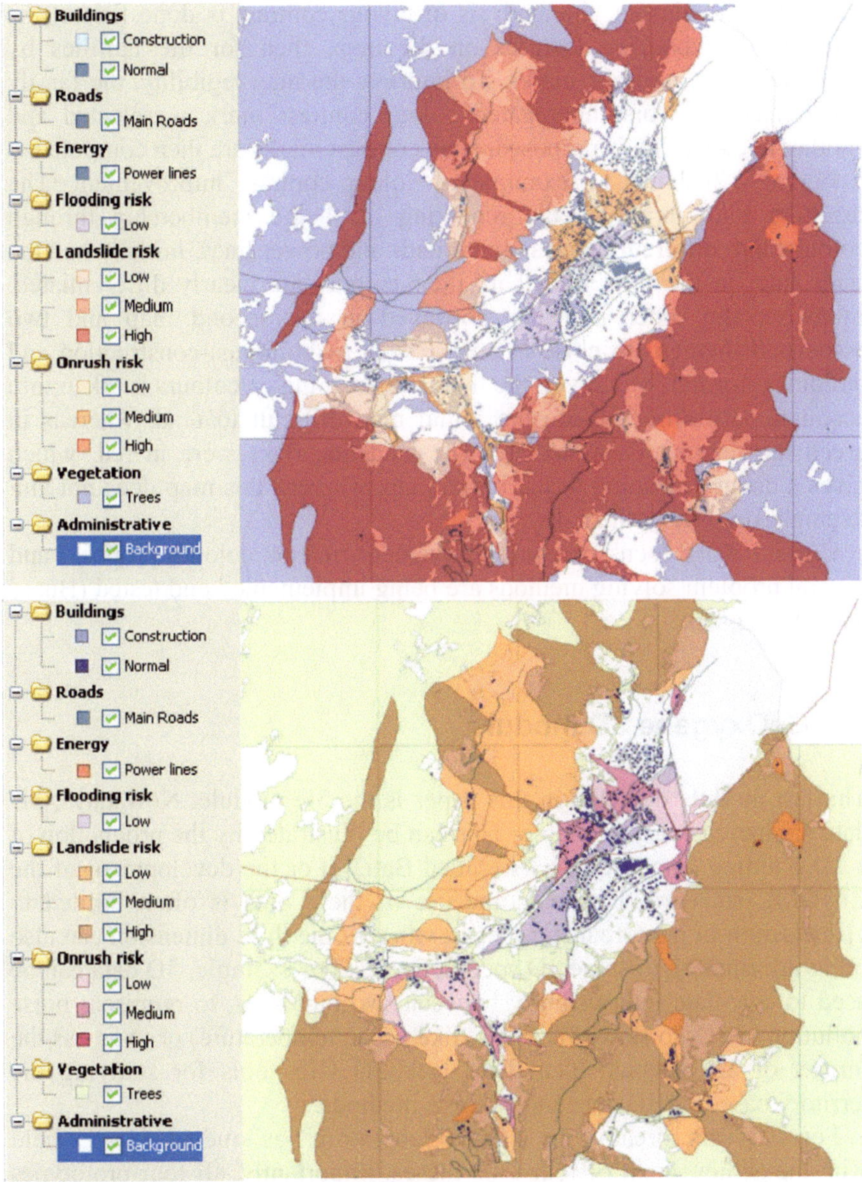

Fig. 2.2. The initial map (*top*) is automatically improved (*bottom*)

The package contains Java classes supporting the definition of a cartographic structure:

- The family of features which are portrayed with the same colour (making up a theme)
- The links between neighbouring features on the map

From these classes, the analysis of colour contrast is done first at the level of neighbouring features on the map, then for the families by aggregating the contrast marks. To improve the map legibility, the family which has the worst aggregated colour contrast mark is selected and another hue or value is proposed. New contrast marks are then computed.

Figure 2.2 shows an example of colour contrast improvement. The resulting map improves the rendering of theme membership through colours: two different themes, main roads and power lines, initially had the same hue; in the improved map, their colours are clearly differentiated. Likewise, the order is better portrayed in the second map and two associated themes are close in colour such as buildings-construction and buildings-normal. Moreover, the thematic meaning of colours is taken into account: purple trees made the initial map difficult to read, whereas in green the reading is immediate. The landslide risks were in red, which gives a danger meaning to the initial map, whereas this map does not aim to emphasise one of the risks.

Current work focus on the improvement of new colours proposal and several problem solving methods are being implemented and tested (Buard and Ruas 2009).

2.5 GeOxygene 3D module

The last module depicted in this paper is the 3D module. Nowadays, 3D data are increasingly available. This can be illustrated by the production of a 3D building database at IGN (called Bati3D) or the development of the CityGML standard. However, the use of these data is often limited to visualisation in photorealistic scenes; whereas the third dimension can also bring information to solve 3D specific issues. For example, 3D data can be used to calculate intervisibility between two buildings, to represent noise pollution in 3D, to study 3D fields (like ocean temperature) or to assess the impact of an antenna. Yet, there lack 3D GIS tools for studying the territory based on 3D data and analysis methods.

For 10 years already, the COGIT laboratory has studied the 3D field with the achievement of four PhD theses. Importantly, all four prototypes were implemented on different platforms.

- In the first one, De La Losa (2000) proposed a 3D topological model in order to study relationships between objects. He also discussed the representation of a 3D environment in a GIS from 2D datasets.

- Ramos (2002) designed a prototype of 3D GIS applied to the military field. It enables to calculate the shortest path on a DTM and provides intervisibility tools. As developments were led for the army, the laboratory could not keep the code, but only the description of the algorithms used during the thesis.
- Rousseaux (2003) provided methods to assess DTM errors; he demonstrated how to enhance a DTM with topographic data and showed the influence of errors in flood simulation.
- Poupeau and Bonin (2006) developed a 3D GIS prototype named Cristage. It uses crystallographic concepts to provide operations such as building simplification, symmetries detection or relationship studies.

The idea of developing a 3D module on GeOxygene was motivated by two observations. Firstly, the extendibility of the geometric schema of the platform enables the development of 3D modelling and methods. Secondly, the platform will permit to integrate these 3D works and offer the possibility to extend 2D research results to 3D.

To add 3D capacities to GeOxygene, some enhancements were made in different components.

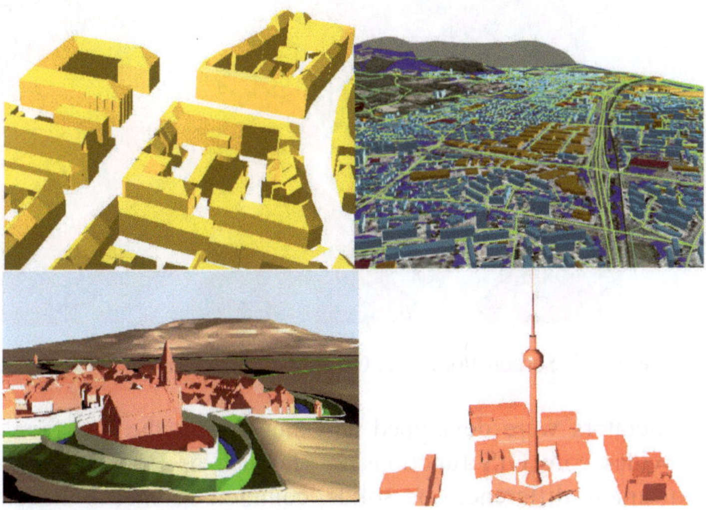

Fig. 2.3. Screenshots from GeOxygene. From *left* to *right* and from *top* to *bottom*: a Bati3D dataset, ESRI shapefiles and DEM data, and 2 City GML datasets

The first step was the improvement of the geometrical model in accordance with the ISO/OGC models (ISO 2003). Some 3D geometric classes were implemented or improved because their existence was unnecessary (for example, the GM_Solid class) or inappropriate (the GM_Surface class) in

2D. To handle 3D data, specific loaders have been implemented to handle different formats (Bati3D, CityGML, Shapefile, PostGIS).

The screenshots in Figs. 2.3 and 2.4 are taken from the platform viewer. A specific interface similar to standard GIS interfaces was developed. This interface manages 3D information using a Java3D scene graph. Layer management, selection tools, light parameters and real-time navigation are provided. Comparatively a 2D standard GIS, the main difference the use of walkthrough navigation.

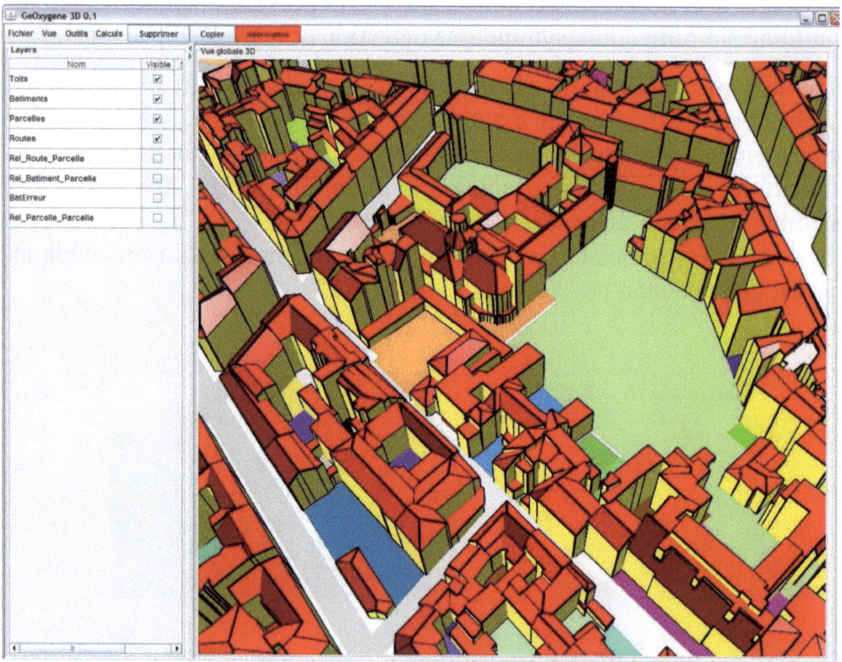

Fig. 2.4. Screenshot of the GeOxygene 3D interface

3D GIS operators were developed and included in the interface. The platform provides core GIS functions in 3D such as Boolean operators (intersection, union, difference) on solids, common geometric algorithms (gravity centre, area, volume, convex hull, …) and solid decomposition (surface triangulation, tetrahedrization).

The first implementation on GeOxygene 3D module was a fast 3D simplification algorithm. It has also been developed to check the validity of the platform. 3D simplification is important in order to decrease the number of polygons in a solid. It is used to store lighter data, to represent scenes faster with fewer polygons or to communicate 3D data by Web services. Our implementation is a variant of Kada (2007). Simplification is made by merging the buffers of the different faces according to a threshold (fusion parameter). Facets of the simplified building are reconstructed by approximating buffers by the median plan. Several levels of simplification are available by changing the value of fusion parameter, as illustrated on Fig. 2.5.

The 3D module of GeOxygene is currently in a finalization phase (the release is planned for the end of 2010). Some improvements are being led in order to ensure the robustness and the completeness of the geometrical operators (offsetting calculation, intervisibility, etc.). Some developments according to territory management are in progress; the first one deals with calculating and representing the optimal hull of buildings according to regulation constraints. Memory consumption and computation time are more important than 2D operations and can currently cause some problems. Indeed, some difficulties appear when processing on 3D geometries because the application requires solids which are topologically valid (i.e. closed, no intersecting faces etc.).

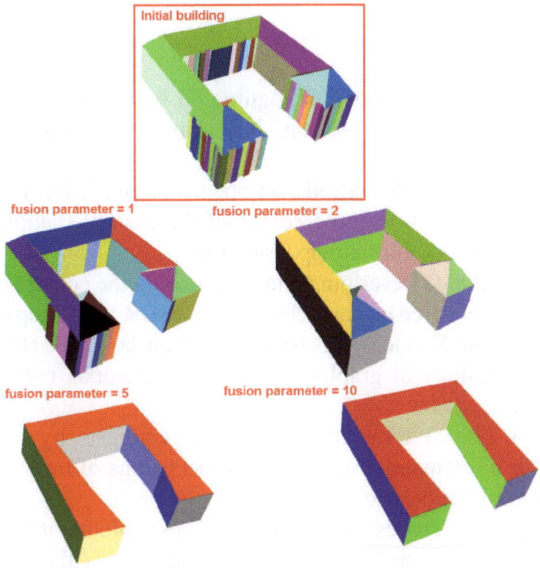

Fig. 2.5. Different simplification results obtained by changing the z-fusion parameter. Each face of the building is represented with a random colour to better visualize the number of faces

2.6 Conclusion

This chapter has presented GeOxygene, a specific result and tool for COGIT research. GeOxygene's specificity is to facilitate the loading of geographical data into object oriented data models, namely into Java structures. This has allowed the development of expert modules that synthesise a long expertise in GI Science (IGN expertise) and years of research such as the data matching module and the colour contrast management module. These modules are implementations of IGN expertise. Yet, to our mind, GeOxygene is more than that. It is a central tool to support important research activities like capitalisation, collaboration, dissemination and integration. The 3D module is for example an achievement regarding capitalisation and also regarding the transfer of 2D knowledge to 3D. GeOxygene can also become a medium to transfer knowledge from IGN to the community of developers eager to develop innovative GIS applications.

References

Abd El Kader Y, Bucher B (2006) Cataloguing GI functions provided by non web services software resources within IGN. In: Proceedings of the AGILE conference, Budapest

Badard T, Braun A (2003) OXYGENE: an open framework for the deployment of geographic web services. In: Proceedings of the 21st international cartographic conference (ACI/ICA), Durban, pp 994–1004

Balley S, Gesbert N (2004) Describing and customising the content of geographical datasets: approaches based on data schemas. EuroSDR Workshop on Ontologies & Schema Translation Services, 15–16 avril, Marne-la-Vallée (France)

Bertin J (1983) Semiology of graphics: diagrams, networks, maps. University of Wisconsin Press, Madison

Buard E, Ruas A (2007) Evaluation of colour contrasts by means of expert knowledge for on-demand mapping. In: Proceedings of the 23rd international cartographic conference, Moscow

Buard E, Ruas A (2009) Processes for improving the colours of topographic maps in the context of Map-on-Demand. In: Proceedings of the 24th international cartographic conference, November, Santiago

Bucher B, Jolivet L, Buard E, Ruas A (2007) The need for web legend services. In: Ware JM, Taylor GE (eds), In: Proceedings of the 7th international symposium on web and wireless GIS (W2GIS), Springer Lecture note in computer science, Cardiff

Chesneau E, Ruas A, Bonin, O (2005) Colour contrasts analysis for a better legibility of graphic signs on risk maps. In: Proceedings of the international cartography conference ICC' 2005. La Coruna

De La Losa A (2000) Modélisation de la troisième dimension dans les bases de données géographiques, PhD thesis in GI Science of the University of Marne La Vallée

Grosso E, Bouju A, Mustière S (2009) Data integration geoService: a first proposed approach using historical geographic data. In: Carswell JD, Fotheringham AS, McArdle G (eds), In: Proceedings of the web and wireless geographical information systems, 9th international symposium, W2GIS 2009, Springer, Maynooth, 7–8 Dec 2009, pp 103–119

Hampe M, Sester M (2002) Real-time integration and generalization of spatial data for mobile applications. Geowissenschaftliche Mitteilungen, Maps and the Internet, Wien, Heft (60):167–175.

ISO (2003) Geographic information – Spatial Schema ISO Draft International Standard 19107:2003

Kada M (2007) 3D building generalisation by roof simplification and typification. In: Proceedings of ICC2007, the international cartography association conference, Moscow

Kilpelaïnen T (2000) Maintenance of multiple representation databases of topographic data. Cartogr J 37(2):101–107

Mustière S, Devogele D (2008) Matching networks with different levels of detail. GeoInformatica 12(4):435–453

Mustière S, Van Smaalen J (2007) Chapter 6: database requirements for generalisation and multiple representations. In: Mackaness W, Ruas A, Sarjakoski T (eds) The generalisation of geographic information: models and applications. Elsevier, London, pp 113–136

Olteanu-Raimond AM, Mustière S (2008) Data matching – a matter of belief. In: Proceedings of the international symposium on spatial data handling SDH'2008, Montpellier, pp 501–519

Perret J, Boffet MA, Ruas A (2009) Understanding urban dynamics: the use of vector topographic databases and the creation of spatio-temporal databases. In: Proceedings of the 24th international cartography conference (ICC'09), Santiago, 15–21 Nov

Poupeau B, Bonin O (2006) Cristage: a 3D GIS with a logical crystallographic layer to enable complex analysis, 3DGeoInfo conference, Kuala Lumpur, 7–8 Aug

Ramos F (2002) A multi-level approach for 3D modeling in geographical information systems, ISPRS Commission IV symposium, Ottawa, 9–12 July

Rousseaux F (2003) Enrichment of a TIN with 3D vectors data, 11th conference on GIScience and research in UK (GISRUK'03), London, 9–11 April

Chapter 3 Use of Open-Source GIS for the Pre-processing of Distributed Hydrological Models

Flora Branger[1], Sonja Jankowfsky[1], Olivier Vannier[2], Pierre Viallet[3], Samuel Debionne[3], Isabelle Braud[1]

[1]Cemagref, UR HHLY, 3bis quai Chauveau CP 220, 69336 Lyon Cedex 09, France {flora.branger, sonja.jankowfsky, isabelle.braud} @cemagref.fr
[2]LTHE, Université de Grenoble (CNRS, INPG, IRD, UJF), BP 53, 38400 Grenoble Cedex 9, France, olivier.vannier@ujf-grenoble.fr
[3]Hydrowide, 1025 Rue de la Piscine, 38600 Saint-Martin d'Hères, France, samuel.debionne@hydrowide.com

Abstract. In distributed hydrological models, elementary modelling units are often derived from raster and vector geographical maps such as digital elevation models (DEM), land use and soil properties maps. Model pre-processing can be defined as the operations transforming raw GIS data into the adequate modelling units. This paper describes our experience with open-source GIS for these pre-processing operations through a set of examples. The softwares tested were PostGIS, OpenJump, OrbisGIS, GRASS, SAGA, MapWindow and QuantumGIS. The pre-processing operations dealt with DEM processing, topology corrections, and advanced editing of vector layers. None of the tested software provided a ready to use solution. However GRASS GIS appeared as the most promising candidate for performing all the preprocessing operations and automating them. Although sometimes tricky to handle by GIS non-expert users, open-source GIS allow easy tests and comparisons and therefore are well suited to research purposes.

E. Bocher and M. Neteler (eds.), *Geospatial Free and Open Source Software in the 21st Century*, Lecture Notes in Geoinformation and Cartography,
DOI 10.1007/978-3-642-10595-1_3,

3.1 Introduction

Hydrological simulation models are valuable tools for addressing issues related to catchment management, as for example the impact of climate change and human activity on water resources, floods, droughts and pollution. Distributed hydrological models are particularly well adapted to these issues, because they can take into account explicitly catchment heterogeneity. In distributed hydrological models, the catchment is discretized into homogeneous spatial units, that form the elementary modelling units over which process equations are calculated. Grid-based models and vector-based models can be distinguished. Grid-based models use the pixels of a raster grid (usually Digital Elevation Model) as elementary spatial units, on which parameter values and boundary conditions are defined. Vector-based models follow a different concept for spatial discretization: the spatial elementary units are irregular vector geometries, called hydrological response units (Flügel 1995), or hydro-landscapes (Dehotin and Braud 2008). They are obtained by intersection and combination of raster and vector maps such as topography, land use, soil properties, or geology. They also depend on the model objectives and data availability. They form very irregular geometries mixing polygon, line or point elements. Using such irregular spatial units has the advantage of representing hydrological discontinuities, which have an important impact on hydrological processes in a catchment despite their small sizes (Moussa et al. 2002): roads, sewer networks, ditches, hedgerows, etc. Many distributed hydrological models use this vector-based spatial discretization (Flügel 1995; Dehotin and Braud 2008; Arnold and Fohrer 2005; Krause et al. 2006). We have been developing such models for several years, with the help of the LIQUID modelling framework (Viallet et al. 2006; Branger et al. 2010).

The delineation of the spatial units for a given model application depends on many factors, such as the objectives of the study, the scale, the model specifications and the availability of data. Some features might be important for a specific application and not for another one (Blöschl and Sivapalan 1995). This issue was addressed by Dehotin and Braud (2008), who proposed a methodology for the delineation of hydro-landscapes and illustrations for several application cases. In addition to their methodological considerations, the present paper deals with rather practical objectives. During the process of hydro-landscape delineation, the original input data must be transformed, overlaid, etc using GIS software packages. This set of operations is called pre-processing.

This paper presents and discusses our experience with free and open-source GIS for these pre-processing operations, on the basis of our developments of distributed hydrological models with LIQUID. At first, the typical application cases and input data on which we built our results will be briefly presented, as well as the GIS software packages that we used (Sect. 3.2). Section 3.3 will present a selection of examples of pre-processing operations and the problems that were encountered. These results will be discussed in Sect. 3.4.

3.2 Material and Method

3.2.1 Application Cases and Data

The examples presented in this paper are taken from three application cases. The first case is the simulation of flash floods on several semi-mountainous catchments, with sizes ranging from 1 to 100 km^2 (Manus et al. 2009) The second application case deals with the modelling of the influence of landscape and water management practices, such as subsurface drainage, ditches, hedgerows or grassed strips, on the hydrology of a small (1.28 km^2) agricultural catchment (Branger et al. 2008). The third application case is included in the AVuPUR research project (Braud et al. 2010), which aims at enhancing the understanding and modelling capacity of water fluxes in suburban watersheds. The objective is to build a model of two small (5 km^2) catchments in order to assess the impact of extending urban zones, sewer networks, ditches and roads on the hydrological regime of the rivers. In the first application case, the hydro-landscapes are the combination of the sub-catchments and of the cartographic soil units taken from a soil properties database. In the other examples, the hydro-landscapes are mainly derived from land-use maps. In all examples, equations and water balance are calculated on these vector spatial units. Therefore the hydro-landscapes must fit perfectly together, like the pieces of a jigsaw puzzle. In particular, no overlapping, no spaces in and between the different polygon elements can be tolerated.

GIS data used for these applications were first of all digital elevation models from various sources: Alti and Topo databases from IGN (French National Geographical Institute), or LIDAR surveys. The input formats were raster grids with resolutions from 2 to 75 m. Maps of the river network, when available, were vector line maps, extracted from the Carthage (French national reference) or Topo databases, or obtained through field measurements. Input data for geology, pedology, and land use were polygon vector maps. Geology maps were digitized from paper maps; pedology maps were obtained as shapefiles from regional agencies for agriculture. For land use, the maps were obtained from the Corine Land Cover European data base, from mapping from aerial photographs (Béal et al. 2009), or from digitizing of field surveys. All these data were gathered from various sources, and were initially produced for contrasted purposes: general cartography over large areas, administration and planning of agricultural activity, etc. This resulted in variable quality of the data, in particular concerning the topology of vector layers.

3.2.2 Software Packages

We used a set of free and open-source GIS software packages under the Microsoft Windows operating system. The list is given in Table 3.1. Except for the PostGIS extension of the PostgreSQL database management system, all of these software systems have been thoroughly reviewed by Steiniger and Bocher (2009) and will not be described in this paper. The choice of softwares was done through a trial and error progression. Historically, we firstly used PostgreSQL/PostGIS, because PostgreSQL/PostGIS tables are the input data sources for most of the models developed in LIQUID (Viallet et al. 2006; Branger et al. 2010). The pre-processing operations were coded as SQL queries using PostGIS functions. OpenJump and QuantumGIS were used as visualization tools because they allowed an easy display of PostGIS tables and had user-friendly interfaces. However, PostGIS did not provide all the functionalities that were needed, in particular for DEM analysis and topology. Therefore we progressively developed a strategy using combinations of several softwares: SAGA GIS and MapWindow for DEM processing; OpenJump, OrbisGIS, QuantumGIS and GRASS for vector topology.

Table 3.1. GIS software packages used for the pre-processing operations

Table	Version	Comment
PostGIS	1.2.1	With PostgreSQL 8.2
OpenJump	The Merge	No additional plugins
OrbisGIS	2.1	–
GRASS	6.4	Windows native version
SAGA GIS	1.2	–
MapWindow	4.5	Test of Taudem toolbox only
QuantumGIS	1.1	Use of plugin fTools 0.5.9

3.3 Pre-processing Operations and Examples of Encountered Problems

This section presents the main pre-processing operations that had to be performed for our application cases. The two first operations (delineation of sub-catchments and correction of the topology of vector layers) are quite classical. They are usually performed for any model dealing with hydrological response units or hydro-landscapes (Lagacherie et al. 2010; Schwartze 2008). The pre-processing operations presented in the last subsection deal with advanced editing of vector layers. They are more specific to the very detailed model applications presented in Sect. 3.2 (Branger et al. 2008; Braud et al. 2010).

3.3.1 Delineation of Sub-catchments

The first pre-processing operation consists of sub-catchment delineation from Digital Elevation Models in the area of interest. For this operation, we tested SAGA GIS and the Taudem toolbox in MapWindow (Terrain Analysis Using Digital Elevation Models, Tarboton 2002). The classical steps for this operation are sink removal, computation of flow directions, slopes, drained areas, river network delineation and sub-catchments delineation using a drained area threshold. We used similar options in SAGA and Taudem, in particular the D8 method for flow directions in both cases. On our small to medium scale data, we observed that some catchments were missed by SAGA, and that unrealistic sub-catchments were often generated at river confluences (Fig. 3.1a). They are not suitable to hydrological modelling without tiresome manual editing. Taudem generated more reliable sub-catchments (Fig. 3.1b) and was therefore preferred. Other solutions, that were not tested here, are also possible, such as the use of a multi flow direction method instead of D8, or the use of other softwares (OrbisGIS, GRASS).

3.3.2 Correction of the Topology of Vector Layers

As already mentioned, the quality of input vector data is variable regarding topology. Polygon layers such as land use maps may contain self-intersecting polygons, overlaps, holes in polygons or spaces between polygons. For line layers such as hydrographic network maps, the main problem is node connection at junctions (river confluences). Only a few

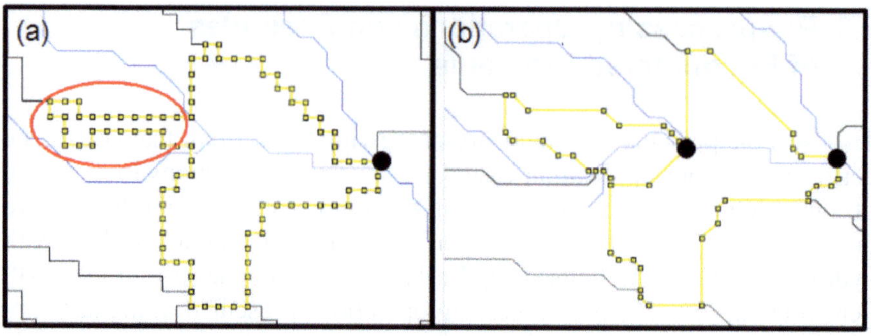

Fig. 3.1. Example of sub-catchments obtained by: (**a**) SAGA. A river confluence is included in the sub-catchment; the part in the circle is incorrect (not drained by the main reach); (**b**) Taudem toolbox in MapWindow. There is one sub-catchment per reach, consistently with river confluences. The sub-catchment outlets are represented with black dots

open-source GIS softwares have topology functions (Steiniger and Bocher 2009). In order to detect and correct topological problems on polygon layers, we tested OrbisGIS, OpenJump and GRASS. The Fig. 3.2 presents the results for a sample of a land use map. OrbisGIS has an *overlapping features* function that generates maps of overlapping polygons, overlapping lines and overlapping points. When applied to our data set, this function detected nearly all the polygons of the layer as overlapping, whether they were actually overlapping or not (Fig. 3.2a). Most of polygon boundaries were also detected as overlapping lines. OpenJump has several quality control functions (in the Quality Control Menu/Detect Overlaps and Validate Layer), that detect overlaps, self-intersecting polygons and holes in polygons. These polygons were correctly detected in our dataset (Fig. 3.2b). However, nor OrbisGIS or OpenJump were able to detect the spaces between polygons. Another drawback is that, up to now, these softwares have no function to correct the detected problems: the layers must be edited manually, which is impossible for large datasets.

On the other hand, GRASS is a topological software that has a fundamentally different model for vector data. In particular, in the case of two adjacent polygons, the common border is stored only once and shared between both areas. Therefore GRASS automatically builds topology and detects problems when loading the input files (*v.build* function). In our data set, the detected gaps and overlaps could be merged with adjacent polygons automatically through the *v.clean* (*rmarea*) function with a threshold of 50 m² (Fig. 3.2d).

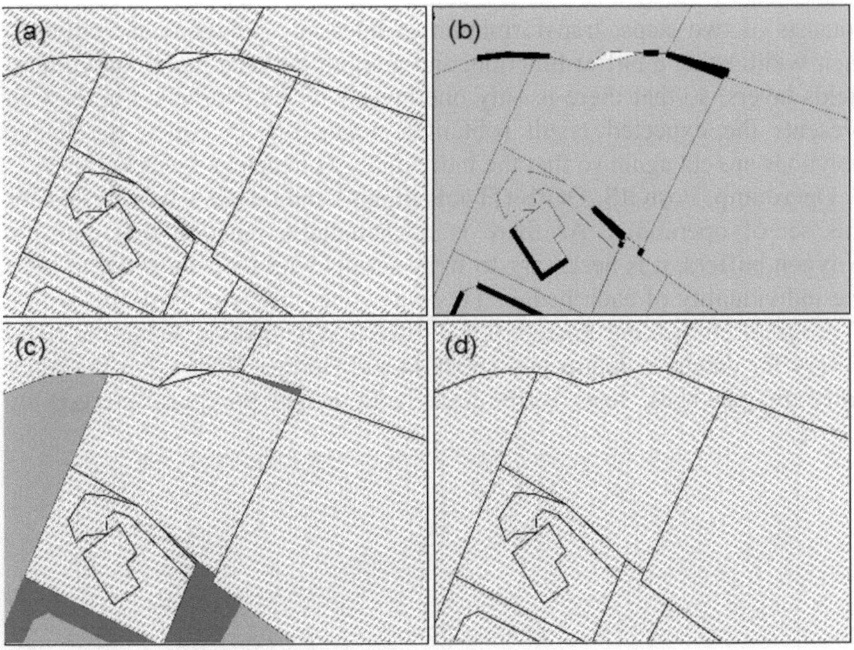

Fig. 3.2. Example of topology processing on a polygon layer: (**a**) Original data; (**b**) OrbisGIS output; overlapping polygons are represented in *gray* and overlapping lines in *black* (**c**) OpenJump output; overlapping polygons are represented in *dark gray* and polygons with holes in *light gray*; (**d**) GRASS output after processing with *v.build* and *v.clean* functions

3.3.3 Advanced Editing of Vector Layers

In addition to the previous pre-processing operations, which must be performed in any case, this section presents our experience with advanced geometrical transformations. These transformations depend on the input data and on the exact requirements of the hydrological model. Therefore all of them are not always necessary.

3.3.3.1 Merging of Several Layers

The merging of several layers may be necessary when the model needs geographical information in a different format than that provided in the input data. Figure 3.3a presents a typical example: the limits of agricultural fields are stored as polygons in one layer, and the positions of hedgerows are stored as lines in another layer. In our model, hedges must be represented as polygons, mainly for water balance concern. The pre-processing operation consists of two steps: transformation of lines into polygons according to their width using a buffer function, and merging of hedges and agricultural fields layers, so that there is only one layer with no overlaps. Figure 3.3b presents the expected result (obtained manually), in which the hedge polygons are cut again so that the individuality of each hedge is kept.

OpenJump, PostGIS, QGIS (fTools plugin) and GRASS were tested for this set of operations. As there is no automatic function that cuts the polygon buffers, it is necessary to write scripts with loops in order to keep the individuality of each hedge. This was done only with PostGIS, with a C program embedding PostgreSQL/PostGIS SQL queries, using the *libpq* library for database connection (Spatola and Branger 2006). For the other software packages, tests were made with hedge polygons merged by

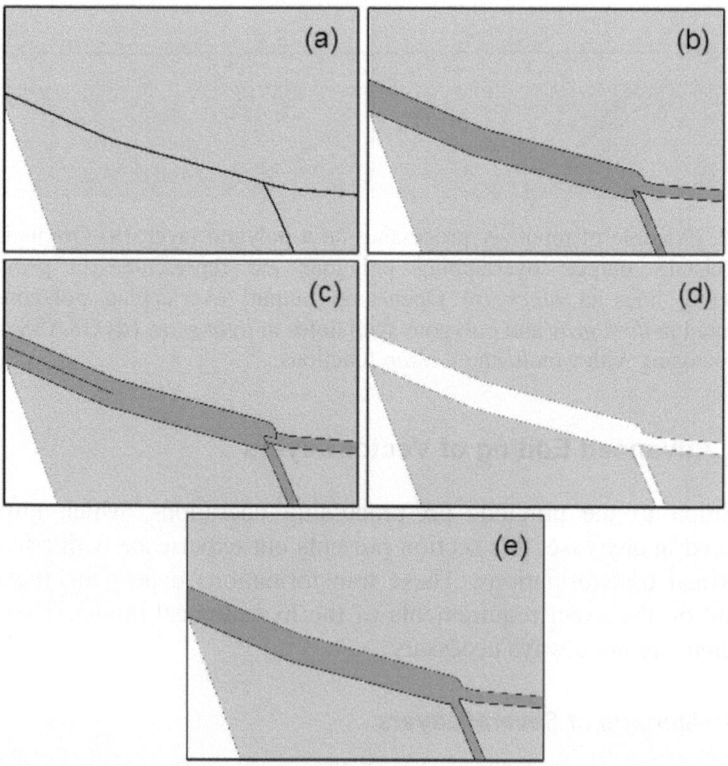

Fig. 3.3. Merging of several layers. The agricultural fields are represented in *light gray*, and the hedges in *dark gray*. (**a**) Original data with hedgerows as lines (in *black*); (**b**) Expected output with individual hedge polygons; (**c**) PostGIS output with a flat, self-intersecting part in the hedge polygon (seen here as a line); (**d**) QGIS output with missing hedge polygons; (**e**) GRASS output

groups for a quick comparison. The functions used were *buffer*, *difference* and *geomunion* (*v.overlay* (*not/or*) in GRASS).

The results are presented in Figs. 3.3c, d, and e. It was not possible to complete the test with our version of OpenJump because the *difference* function works with files containing only one polygon. With PostGIS, we experienced many program crashes, all due to geometry processing errors from the GEOS library at the difference and geomunion steps (generation of invalid geometries at one step and crash at the following step). When not crashing, the combination of PostGIS function generated dirty geometries as shown in Fig. 3.3c. Crashes were also experienced with QGIS, resulting in incomplete output layers (Fig. 3.3d). On the other hand, the combination of GRASS functions was more robust and lead to satisfactory results (Fig. 3.3e), except for the individuality of each hedge. This last point can be resolved by introducing a loop.

3.3.3.2 Intersection of Lines and Polygons

This operation consists of splitting polygons from a layer according to lines from another layer, for example land use polygons according to the river network, as presented in Fig. 3.4. Our detailed hydrological models require a polygon on each border of river reaches, including at river sources. In the same way, the river must be split at each intersection with the adjacent polygons. The input river network and land use maps rarely respect this condition.

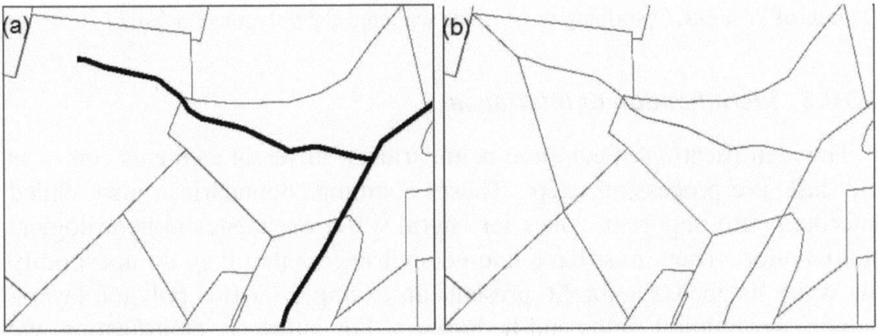

Fig. 3.4. Splitting of land use polygons according to hydrographic network lines: (**a**) Original data; (**b**) Expected output. The *gray* lines are the polygon boundaries and the *thick black* line is the river

Intersection functions of PostGIS, OpenJump, and QGIS, all deriving from the GEOS / JTS geometry processing library, were tested unsuccessfully (Fig. 3.5). They are not operant on line-polygon intersections. We are currently testing GRASS functions with promising results (*v.distance, v.patch*).

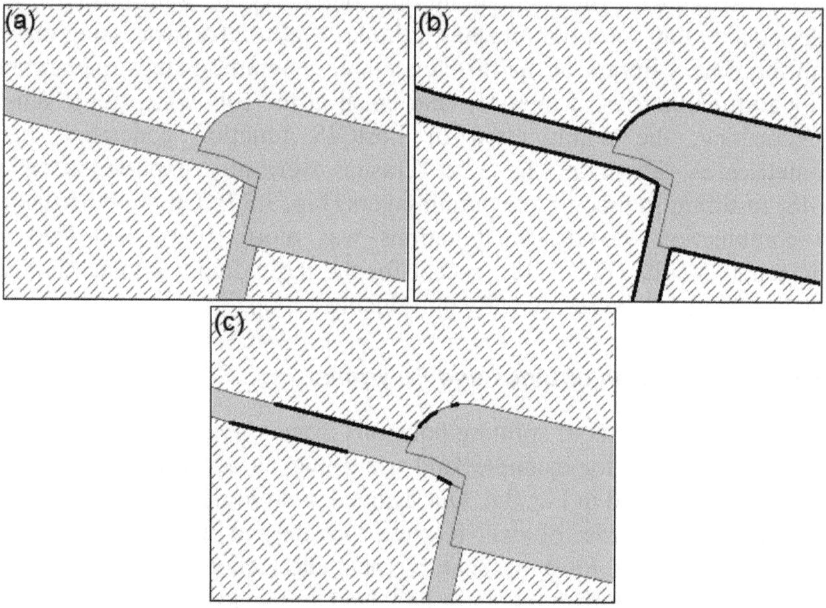

Fig. 3.5. Intersection of two polygon layers for identification of common boundaries: (**a**) Original data (in *gray*); (**b**) Expected output as continuous lines (in *black*); (**c**) Outputs of PostGIS, OpenJump or QGIS as scattered flat polygons (in *black*)

3.3.3.3 Identification of Interfaces

The identification of common boundaries of different elements comes as the last pre-processing step. These common boundaries, also called interfaces, are important zones for lateral water exchanges in hydrological computation. They must have line geometries, so that they do not modify the water balance. Figure 3.6 presents an example for two polygon layers, namely agricultural fields and hedgerows. For interface identification, we used alternatively intersection functions in PostGIS, OpenJump and QGIS. The results were identical in all cases, as shown in Fig. 3.6c, which is not surprising as these software packages use the same geometry processing library (GEOS/ JTS). Instead of lines, the intersection function generated scattered multipolygons or geometry collections, that contained flat, self intersecting polygons, following only partially the element boundaries.

Automatic editing was unsuccessful at transforming these polygons in the expected lines. The only solution was manual editing.

3.4 Discussion

From the test cases described in the previous section, we can see that the pre-processing of distributed hydrological models can be a complex issue that involves a broad spectrum of geographical processing tools, in particular for vector features. Although the set of test cases presented here is quite reduced, it is sufficiently representative to show that several problems were encountered. Among the large number of available open-source tools, only a few were tested in this study. Moreover, not all software packages were used for all test cases. We are aware of this limitation. As stated before, the choice of software was the result of a trial and error progression. Further tests of other tools or new updates would be welcome to complement our results. Yet a few conclusions still can be drawn.

The first conclusion is that a fully topological software such as GRASS is more convenient for processing vector data. In our tests (see in particular Sects. 3.3.2 and 3.3.3.1.), the use of topological/quality check functions in non-topological GIS did not give satisfactory results. It might be possible to improve them but this would probably require several steps and combinations of functions that are beyond our non-GIS specialist skills. Paradoxically, the use of GRASS appeared to be easier. The second conclusion is that some of the processing operations require scripting facilities (see Sect. 3.3.3.1). Scripting is also required in order to automate the pre-processing as far as possible (which is absolutely necessary for large data sets).

At the moment, GRASS appears to be the most promising open-source tool for our needs. The raster and vector processing functions available in GRASS can perform all the operations described in this paper, including topology verification and cleaning. GRASS also provides powerful scripting facilities. In our case, the recent release of a Windows native version and the possibility to write portable Python scripts are advantageous as our models run under the Windows operating system. The recent developments of Schwartze (2008) and Lagacherie et al. (2010) show that GRASS GIS is increasingly used for the pre-processing of hydrological models. This gives us directions for our future developments. The reactivity of the user community forums and mailing lists is also a precious help.

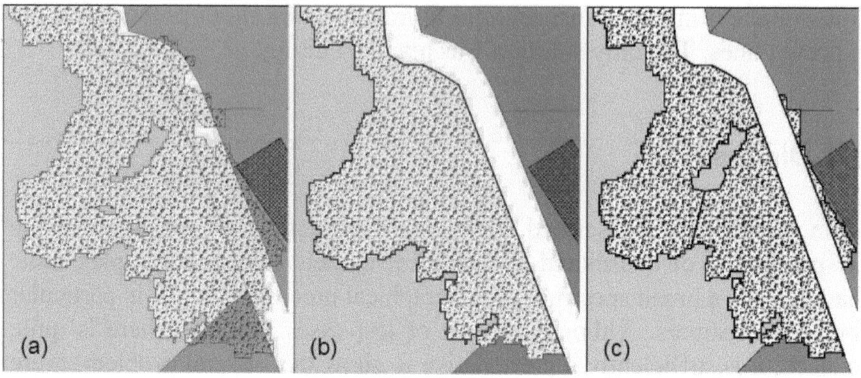

Fig. 3.6. Pre-processing of two layers, representing land use and tree cover. *White color* is for roads, *gray* for agricultural fields, black for buildings, *textured black* for tree cover. (**a**) Original data; (**b**) Simplification with GRASS; (**c**) Ideal simplification minimizing loss of information

But even with the ideal GIS tool, a more general question about pre-processing is how far automation must go. Figure 3.6a provides an example of input overlapping layers. The tree cover layer also has too much detail, with holes and irregular polygon edges. These layers can be merged and simplified with GRASS (Fig. 3.6b) or other GIS. However, some important information are lost in the process. The surface of the tree covered zones is not respected, which may distort the water balance in the model through the calculation of rainfall interception by the trees and tree transpiration. The deletion of the small tree-covered zone on one side of the road may also be a problem. The ideal solution for the merging of several polygon layers is represented in Fig. 3.6c. In this solution, the tree covered zone is divided in several polygons to fulfill topological constraints. The smallest polygons are merged together, with surface exchanges with the other cadastral units, so that the final geometries are simpler while the overall tree covered surface remains unchanged. At the moment, this can be done only manually.

3.5 Conclusion: Interest of Open-Source GIS

Open-source GIS are not always easy to use by GIS non-experts: graphical interfaces may not be as user friendly as that of proprietary products, documentation may be incomplete, the software maintenance may depend on the free time of the developers, etc. For our user profile, that could be defined as "non-expert, but needing advanced tools and functions", the world of open-source GIS could be described as a dense forest where you have to find your way.

However, the big advantage of open-source GIS is that several tools and solutions can be tested and compared easily. Such comparisons would have been impossible with proprietary software, the first reason being license cost. Although they are not experts, researchers need to understand how their tools work and to share their experience as widely as possible. Therefore open-source GIS are well suited for research purposes.

Acknowledgments

The work presented here was mainly funded by the EU FP6 HYDRATE and French ANR AVuPUR projects. We would also like to thank Mélanie Bertrand for helping with the catchment delineation processing.

References

Arnold J, Fohrer N (2005) SWAT2000: current capabilities and research opportunities in applied watershed modelling. Hydrol Process 19(3):563–572

Béal D, Gagnage M, Jacqueminet C, Kermadi S, Michel K, Jankowfsky S, Branger F, Braud I (2009) Cartographie de l'occupation du sol pour la modélisation hydrologique spatialisée du cycle de l'eau en zone péri-urbaine. In: Actes du XXVIIe congrès INFORSID, Atelier SIDE2009 "Systèmes d'Informations et de Décision pour l'Environnement", Toulouse, pp 23–32

Blöschl G, Sivapalan M (1995) Scale issues in hydrological modelling: a review. Hydrol Process 9:251–290

Branger F, Braud I, Viallet P, Debionne S (2008) Modelling the influence of landscape management practices on the hydrology of a small agricultural catchment. In: 8th international conference on HydroScience and engineering, Nagoya, pp 586–594

Branger F, Braud I, Debionne S, Viallet P, Dehotin J, Henine H, Nedelec Y, Anquetin S (2010) Towards multi-scale integrated hydrological models using the LIQUID framework. Overview of the concepts and first application examples. Environ Modell Softw. doi:DOI:10.1016/j.envsoft.2010.06.005

Braud I, Chancibault K, Debionne S, Lipeme Kouyi G, Sarrazin B, Jacqueminet C, Andrieu H, Béal D, Bocher E, Boutaghane H, Branger F, Breil P, Chocat B, Comby J, Dehotin J, Dramais G, Furusho C, Gagnage M, Gonzalez-Sosa E, Grosprêtre L, Honegger A, Jankowfsky S, Joliveau T, Kermadi S, Lagouy M, Leblois E, Martin JY, Mazagol PO, Michel K, Molines N, Mosini ML, Puech C, Renard F, Rodriguez F, Schmitt L, Thollet F, Viallet P (2010) The AVuPUR project (Assessing the Vulnerability of Peri-Urban Rivers): experimental setup, modelling strategy and first results. In: Novatech 2010 conference, 10 pp

Dehotin J, Braud I (2008) Which spatial discretization for distributed hydrological models? Proposition of a methodology and illustration for medium to large-scale catchments. Hydrol Earth Syst Sc 12:769–796

Flügel WA (1995) Delineating hydrological response units by geographical information system analyses for regional hydrological modelling using PRMS/MMS in the drainage basin of the River Bröl, Germany. Hydrol Process 9(3–4):423–436

Krause P, Bäse F, Bende-Michl U, Fink M, Flügel W, Pfenning B (2006) Multiscale investigations in a mesoscale catchment – hydrological modelling in the Gera catchment. Advances in Geosciences 9:53–61

Lagacherie P, Rabotin M, Colin F, Moussa R, Voltz M (2010) Geo-MHYDAS: a landscape discretization tool for distributed hydrological modeling of cultivated areas. Comput Geosci. doi:DOI: 10.1016/j.cageo.2009.12.005

Manus C, Anquetin S, Braud I, Vandervaere JP, Creutin JD, Viallet P, Gaume E (2009) A modeling approach to assess the hydrological response of small mediterranean catchments to the variability of soil characteristics in a context of extreme events. Hydrol Earth Syst Sc 13:79–97

Moussa R, Voltz M, Andrieux P (2002) Effects of the spatial organization of agricultural management on the hydrological behaviour of a farmed catchment during flood events. Hydrol Process 16(2):393–412

Schwartze C (2008) Deriving hydrological response units (HRUs) using a web processing service implementation based on GRASS GIS. In: Proceedings on the workshop geoinformatics FCE CTU 2008

Spatola C, Branger F (2006) Prétraitement automatisé des données pour la réalisation de simulations d'un modèle hydrologique spatialisé. In: Géo-événement 2006, Paris, 16–18 mai 2006

Steiniger S, Bocher E (2009) An overview of current free and open-source desktop GIS developments. Int J Geogr Inf Sci 23(10):1345–1370

Tarboton D (2002) Terrain analysis using Digital ElevationModels (TauDEM). Utah State University. http://hydrology.neng.usu.edu/taudem/Accessed 23 June 2009

Viallet P, Debionne S, Braud I, Dehotin J, Haverkamp R, Saadi Z, Anquetin S, Branger F, Varado N (2006) Towards multi-scale integrated hydrological models using the LIQUID framework. In: 7th international conference on hydroinformatics, nice, vol 1, France, pp 542–549

Chapter 4 A Platform for Spatial Data Labeling in an Urban Context

Julien Lesbegueries[1], Nicolas Lachiche[1], Agnès Braud[1], Grzegorz Skupinski[1], Anne Puissant[1], Julien Perret[2]

[1]LSIIT, Université de Strasbourg, Pôle API, boulevard Sébastien Brant BP 10413 67412 Illkirch Cedex, France {julien.lesbegueries, nicolas.labiche, agnes.braud, grzegorz.skupinki} @lsiit-cnrs.unistra.fr
[2]Institut Géographique National, Laboratoire COGIT 73, avenue de Paris 94160 Saint Mandé, France julien.perret@ign.fr

Abstract. This chapter presents a platform for classifying urban areas, improved by a machine learning framework able to ease this classification. We propose thanks to this platform an iterative procedure for geographic experts that have to define classes or "labels" and then classify in a semi-automated way. This work is part of the GeOpenSim project and has been developed within the Geoxygene framework.

4.1 Introduction

The GeOpenSim project aims at developing an open source framework to study urban evolutions using vector based topographic databases. This framework, based on GeOxygene (Badard and Braun 2003), is composed of several modules spanning from the creation of spatio temporal topographic databases to the simulation of urban dynamics. The novelty of our approach lies in using topographic data for the simulation whereas most research on urban simulation is based on cellular automata or graph cellular automata (Barros 2003; Batty 2005; Benenson and Portugali 1997; Hammam et al. 2007; O'Sullivan 2001). Indeed, the analysis of observed urban phenomena using topographic data allows for a more accurate and more realistic approach to urban simulation. Such an analysis is realized at different geographic levels: at the micro level (buildings, roads, etc.) and at

E. Bocher and M. Neteler (eds.), *Geospatial Free and Open Source Software in the 21st Century*, Lecture Notes in Geoinformation and Cartography,
DOI 10.1007/978-3-642-10595-1_4,

several meso geographic scales (urban blocks, districts, cities, etc.), already used in (Gaffuri and Trévisan 2004; Steiniger and Weibel 2007). For each of these scales, the context in which the evolutions take place is crucial to the study. The context of an evolution is time-based (evolutions are different in the 1950s and in the 1970s) as well as spatially based (evolutions in a peri-urban context are different from the evolutions in an industrial context). Therefore, in the framework of this spatial context, geographic features have to be characterized so they can be labeled depending on their nature. This paper presents our work on the labeling of a specific type of geographic features: elementary areas (or urban blocks). Here labeling refers to the manual assignment of existing class labels to elementary areas chosen by a human expert. A machine learning technique could then use those examples to extract a model to predict the class labels for remaining elementary areas. The proposed labeling platform could therefore be used to automate the classification of urban areas for urban planning and prospective analysis.

We developed an open source add-on to GeOxygene. This labeling add-on fulfills several needs concerning the management of geographic data visualization and labeling. Its input consists in geographic data and a list of labels (Fig. 4.1). Geographic data consists in several layers of topographic data, at micro and meso scales (Fig 4.1a). A target layer is selected in order to define the elements to be labeled (for example, areas in Fig. 4.1c).

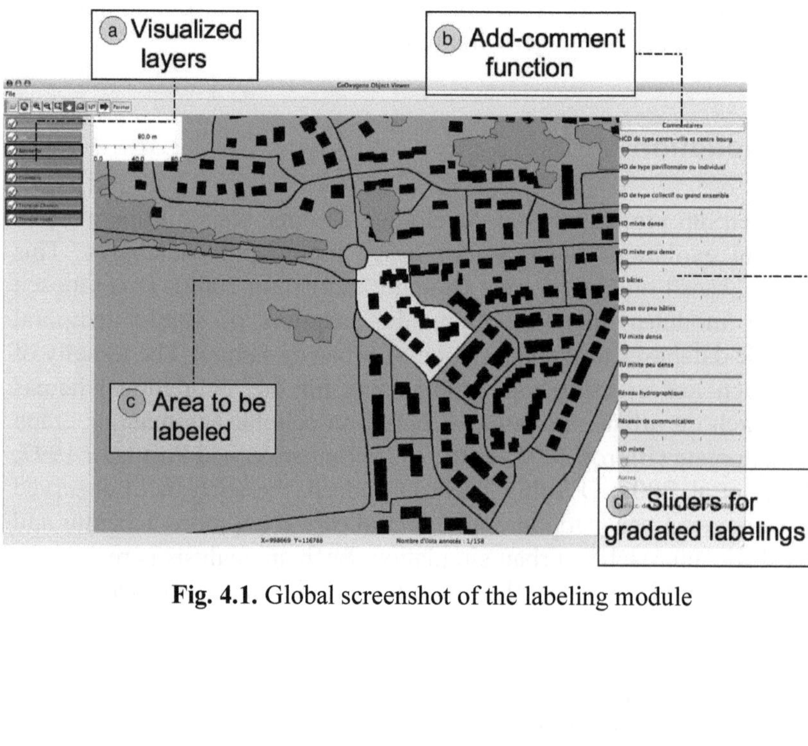

Fig. 4.1. Global screenshot of the labeling module

The other ones are used in the visualization. Moreover a list of labels is defined (Fig. 4.1d) and the module generates a dynamic interface with items or sliders for each label. Sliders allow the user to assign several classes to a single area. If needed, the user can also add a comment (Fig. 4.1b).

The module provides a labeling process with several options in order to perform various experiments and to fit to different problems:

- Several visualizations are provided.
- Several ways of labeling.
- Several procedures to complete the labeling.

The different options are explained and illustrated with the specific problem of urban labeling.

The Fig. 4.2 summarizes the main procedure of labeling. A cartographic window is created from a geographic database with connection functions provided by GeOxygene (Badard and Braun 2003) for specific geographic data. A user logs in to the interface and labels selected areas manually. The labels are stored in a database and export functions produce appropriate output files or output storing in training databases. This data is then used to perform a learning process on the entire dataset. One way is to integrate learning functionalities in our platform with the help of the Open Source Data Mining toolkit Weka (Witten and Frank 2005). This allows us to analyze learning results directly in the platform.

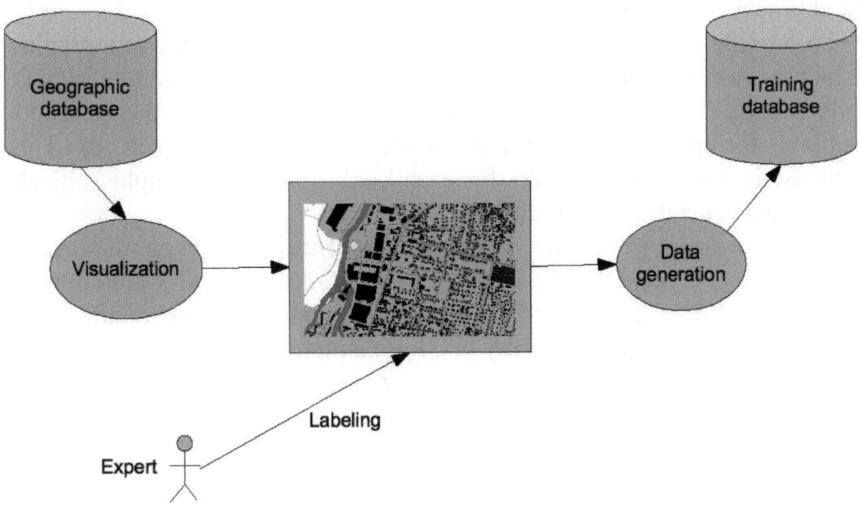

Fig. 4.2. Labeling procedure schema

In the next section, facilities concerning the problem modeling are described: firstly how to carry out the labeling process, and secondly different ways of labeling. Section 4.3 presents the labeling acquisition, its analysis and its use for a supervised learning purpose.

4.2 Modeling the Problem

This section describes how the platform helps modeling the problem. A first step consists in defining the geographic input data and once the geographic parcels are defined, the second step consists in defining the labels.

In the framework of our problem, we must define adequate geographic parcels in a city in order to locate potential evolutions. The first step is to partition it in relevant elementary areas (districts). This partitioning must be generic and automatic. Then labels must be defined to categorize these areas according to their evolution potential. The former function of urban partitioning is performed within GeOxygene and the labeling module performs the latter.

4.2.1 Input Data

Input data must be geographic layers stored in a geographic database e.g. PostGIS.[1] One of the layers must be the target one, in which geographic parcels have to be labeled.

To provide a generic labeling module, we chose to use the topographic database, provided by the French National Geographic Institute (IGN). Indeed there is a vector based topographic database, namely the BDTopo, available for the entire French territory and thus for every French city. This database is composed of layers for each kind of *micro* geographic objects (buildings, vegetation, networks, etc.).

The partitioning of the city is computed according to the communication network layers: main roads, country roads, railroads, and watercourses. Figure 4.3 shows the resulting elementary areas that correspond to the spatial unit to label (the light grey one for example).

[1] http://postgis.refractions.net/

4.2.2 Labels Definition

The platform allows to dynamically define labels, because they can change all along the labeling campaign. A property file configured *a priori* is used to create a label index in a database (Fig. 4.4) and to build the list of sliders (Fig. 4.1c).

Labels are similar to those used in previous studies such as (Steiniger et al. 2008):

1. Continuous urban fabric (city center)
2. Discontinuous urban fabric with individual houses
3. Discontinuous urban fabric with collective buildings
4. High density mixed housing surface (mix of 2 and 3)
5. Low density mixed housing surface (mix of 2 and 3)
6. Specific urban surface (industry buildings)
7. Not or little built specific urban surface (industrial wasteland)
8. High density mixed urban surface (mix of 2,3 and 6)
9. Low density mixed urban surface (mix of 2,3 and 6)
10. Hydrographic network (canals, rivers)
11. Communication network (roads, country roads, railroads)

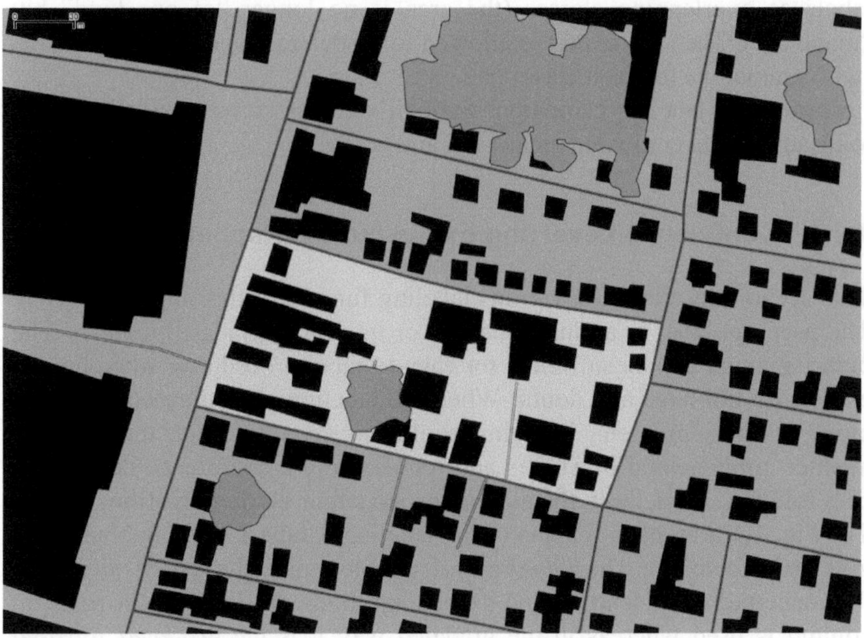

Fig. 4.3. Elementary areas (in *grey* and *light grey*) are built from communication networks (*lines*). The black polygons represent buildings and the rounded areas represent vegetation

For each label, a slider and a tuple in the database are created. This dynamic building is necessary because the labeling procedure is intrinsically an iterative, possibly backtracking procedure. Next section details this characteristic.

4.2.3 Refining the Labels

The module provides different functionalities to refine the labeling. A relevant question is the manner labels are collected.

The easiest way is to force the user (urban planners for instance) using the platform to choose one class label for each visualized area during the labeling process. However, in the case of a random choice of next areas to be labeled, the expert may have difficulties to label every area, given that the areas generation is automatic and can produce artifacts. An additional class "others" can solve the problem.

A second implemented solution consists in allowing a gradation the expert can express with a confidence degree from 1 to 4. In this case the value of each class is stored for each labeled area.

Another manner to face the difficulty of labeling is to consider areas labels as overlapping classes (the area is no longer "of one kind" but "made of"). Then the expert is allowed to associate several classes by area (with gradations for each class).

During our labeling campaign, some mixed classes were added in order to disambiguate confusing areas (4, 5 and 8, 9).

4.2.4 Confidence Level and Exclusive/Overlapping Classes

This section describes the various labeling functionalities implementation. The user can label in a binary manner or in a more gradated manner. The binary manner can be sufficient for easy-to-label procedures, when human experts do not have any doubt. When the labeling procedure consists in a more complex problem, implying confusing areas to label, the gradated manner implemented by sliders and representing a confidence degree can be a solution. By default, the module expects four gradations (from 0 to 3). The Fig. 4.4 presents a schema of the labeling database (storing binary and gradated labelings). The *labeling index* table stores the expert identifier, the concerned area identifier and its labeling. In the binary way, the *label_id* column is used whereas in the gradated way it is the *gradated_label_id* one (on the figure, both columns are filled in order to illustrate the two cases). Then the *Gradated index* stores the sliders values.

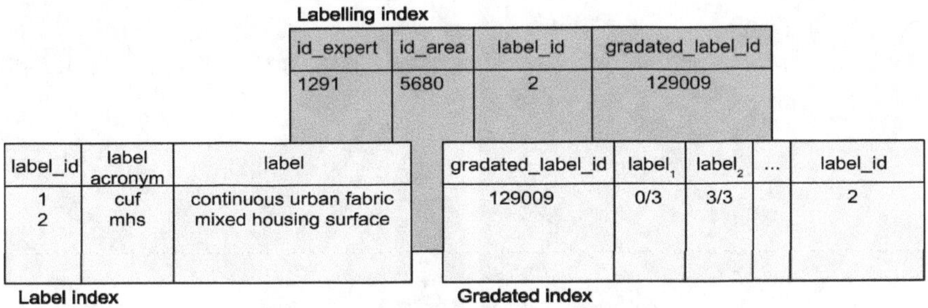

Fig. 4.4. Schema of the labeling database

Moreover, the labeling process can be a one label per area labeling or a several labels per area labeling. It depends on the labels definition. Indeed, they can represent exclusive classes or overlapping classes. In our case study, labels cannot be exclusive because an area can contain *housing surface* and *specific urban surface* at a time. Our problem turns out to be an exclusive labels case, if an additional specific one (*mixed urban surface*) is defined. However the labeling structure can manage the two solutions. The only change is the function used to determine the *label_id* column of the gradated index (max of the $label_i$ columns for instance).

4.2.5 Identifying the Minimum Background

Additionally, the module provides different manners to choose areas to be labeled. Indeed, in a learning perspective, areas have to be well chosen in order to correspond to a representative set. Several solutions are provided (randomly-based, user-based, active learning).

Once input data and classes are correctly defined, we experiment several ways of labeling in order to find the best method requiring the minimum information to display, providing however sufficient information to experts and identifying the adequate retrieval of information necessary for the learning task.

Three visualizations are proposed in our platform (Fig. 4.6), providing different widths of spatial context:

1. The area to label only (with its buildings)
2. The area to label and its surrounding areas (with their micro objects)
3. The entire map of the city

In our case study, experts claim that the first visualization is too poor and there is not enough information to make a decision. For example, it is difficult to distinguish a city center area from a specific urban one (industrial area for instance) without the direct neighborhood and their scale (Fig. 4.5).

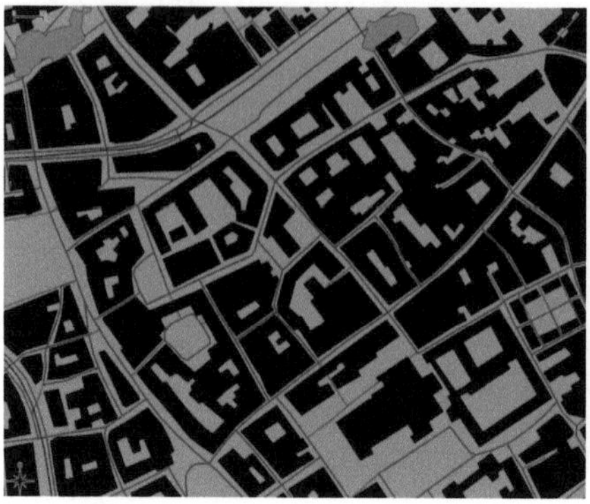

(a) City center

(b) Industrial area (on the left)

Fig. 4.5. Confusing areas which are difficult to label

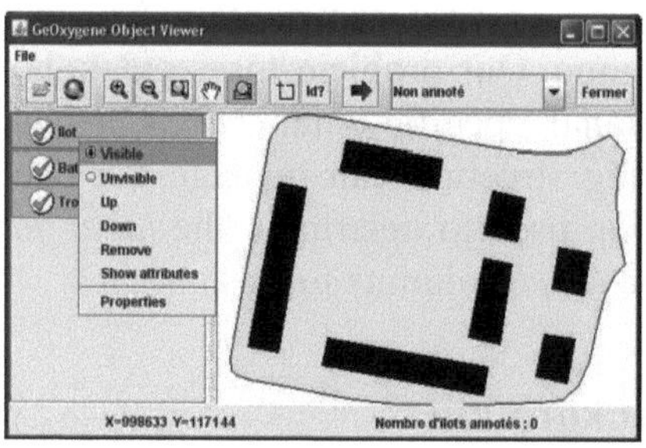

(a)

(b)

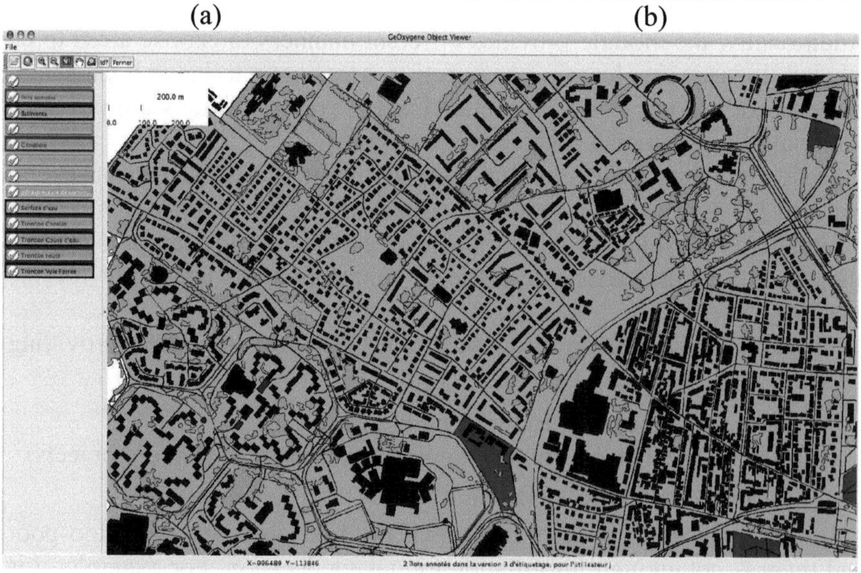

(c)

Fig. 4.6. Three visualization widths offered by the module

The two other visualizations provide this additional contextual information. However, there is a problem of *over-contextualization* for the third one because the visualization of the entire map causes the expert to use his background knowledge of the geographic area (experts could then recognize the displayed city) instead of the conformation of objects in areas.

4.3 Data Processing

This section presents module facilities concerning the automation of the labeling procedure: data acquisition, analysis and visualization, and its use in a learning process.

4.3.1 Data Acquisition

The problem depicted here concerns the manner areas to label are chosen. Several solutions are proposed on the platform. One solution is to visualize a random area not yet labeled (Figs. 4.6a and 4.6b). An alternative solution provided consists in allowing the expert to choose the areas to label (Fig. 4.6c).

The aim of the platform is to include a learning process, able to label the majority of areas from a few examples. With the second solution, experts have to carefully choose the areas. They should not select the ones easy to label only. If the target of the learning process is to label every area, it must be trained from typical and also fuzzier/tricky/complex areas. Another solution would be to use active learning methods (Bondu and Lemaire 2007) to select next areas to label.

4.3.2 Analysis and Visualization of Labelings

This section presents additional functionalities in order to compare labeling between experts. Indeed, it is important to check whether different experts agree on the chosen labeling process: the set of labels and their definition, and the sample of examples.

A particular attention is given to the agreement of labeling between them. First of all, a visualization of the labeling progress is provided for each expert, for each part of the city, in order to indicate to label in priority the areas that have already been labeled by other experts. Then an agreement visualization (Fig. 4.7) allows seeing which areas are the most conflicting ones. The agreement measure is (Eq. 4.1):

$$U = \frac{\sum \sqrt{\sum (m_c - e_i(c))^2}}{p} \qquad (4.1)$$

where

- e_i denotes the experts (there are p experts),
- c denotes the classes (the cardinality of the set of classes is n),
- m_c denotes the average over the different experts of the values for the c class,
- $e_i(c)$ denotes the value chosen by the ith expert for the c class

A greater value of U denotes a stronger disagreement of the experts

(a) Agreement of experts with first settings	(b) Agreement of experts with other settings	(c) Legend

Fig. 4.7. Agreement variation along the experiments

The Fig. 4.7 shows a clear improvement between the first and the latter labeling process. These kinds of analysis allow the experts to choose the best way of labeling.

In order to eliminate residual conflicts, we created a functionality allowing experts to add comments (Fig. 4.1b) to their labeling and to visualize areas by comment (not-well formed area, city border area, etc.).

4.3.3. Towards Automating the Labeling Process

Finally, export functionalities allow using labeling results in learning programs. In particular, exports in arff format for Weka (Witten and Frank 2005), an Open Source learning toolkit, and in 1BC format for 1BC, an

ILP[2] learning tool (Flach and Lachiche 2004) are provided. This is a first step to automate the labeling from a training set. The Fig. 4.8 shows an excerpt of automatically labeled map, produced by a machine learning algorithm applied on experts training sets from the urban labeling procedure.

A part of the Weka library will be integrated in the module in order to envisage a semi-automatic learning process based on a supervised classification. Moreover, this integration allows us to imagine active learning functionalities (Bondu and Lemaire 2007) that could improve the labeling by choosing the most appropriate areas to label for an efficient learning process.

Fig. 4.8. Predictions of a supervised classifier trained from data labeled thanks to the labeling platform

2 Inductive Logic Programming.

4.4 Conclusion

This paper presents the manual labeling functionalities of a GeOxygene[3] extension, within the framework of an urban classification. The module aims at being generic in order to perform similar classifications for other geographic layers. We investigated pertinent facilities for a labeling module, like the labels management, the confidence level capability, the exclusive or overlapping classes choice, the visualizations, and the analysis and use of labeling results.

Future works concern experiments of the integrated learning module, in order to classify the whole dataset and the active learning option allowing the machine to choose the best training examples, i.e. to speed up the collection of training data.

References

Badard T, Braun A (2003) Oxygene – d'une plate-forme interopérable au déploiement de services web géographiques. Revue internationale de géomatique 3(13):411–430

Barros JX (2003) Simulating urban dynamics in Latin American cities. In: Proceedings of the 7th international conference on GeoComputation, University of Southampton, United Kingdom

Batty M (2005) Cities and complexity: understanding cities with cellula automata, agent-based models, and fractals. MIT Press, Cambridge

Benenson I, Portugali J (1997) Agent-based simulations of a city dynamics in a gis environment. In: COSIT '97: Proceedings of the international conference on spatial information theory. Springer-Verlag, London, pp 501–502

Bondu A, Lemaire V (2007) État de l'art sur les méthodes statistiques d'apprentissage actif. RNTI, Numéro spécial dur l'apprentissage et la fouille de données

Flach P, Lachiche N (2004) Naive Bayesian classification of structured data. Mach Learn 57(3):233–269

Gaffuri J, Trévisan J (2004) Role of urban patterns for building generalisation: an application of AGENT, 7th ICA workshop on generalisation and multiple representation, Leicester, 20–21 Aug

[3] GeOxygene aims at providing an open framework which implements OGC/ISO specifications for the development and deployment of geographic (GIS) applications. It is an open source contribution of the COGIT laboratory at the IGN (Institut Géographique National), the French National Mapping Agency. It is released under the terms of the LGPL (GNU Lesser General Public License) license.

Hammam Y, Moore A, Whigham PA (2007) The dynamic geometry of geographical vector agents. Comput Environ Urban 31(5):502–519
O'Sullivan D (2001) Graph-cellular automata: a generalised discrete urban and regional model. Environ Plann B 28:687–705
Steiniger S, Weibel R (2007) Relations among map objects in cartographic generalization. Cartogr Geogr Inform Science (CaGIS) 34(3):175–197
Steiniger S, Lange T, Burghardt D, Weibel R (2008) An approach for the classification of urban building structures based on discriminant analysis techniques. Transactions in GIS 12(1):31–59
Witten IH, Frank E (2005) Data mining: practical machine learning tools and techniques, 2nd edn. Morgan Kaufmann. San Francisco, United States of America

Chapter 5 TAnaTo2: A Tool to Evaluate the Impact of Natural and Anthropogenic Artefacts with a TIN-Based Model

Erwan Bocher, Jean-Yves Martin

IRSTV – FR CNRS 2488 Ecole Centrale de Nantes, 1 rue de la Noë 44321 Nantes Cedex 3, France {erwan.bocher, jean-yves.martin} @ec-nantes.fr

Abstract. This paper presents the TAnaTo2 model. It is based on the open source GIS OrbisGIS, developed at IRSTV institute. TAnaTo2 provides new methods to evaluate the impact of natural and anthropogenic artefacts on the water runoff pathways in a peri-urban basin with a TIN-based model. It allows you to build a drainage network which represents runoff direction computed from the steepest maximum down slope. It is used to pinpoint water flow accumulation and extract watersheds. Tanato2 is applied on the Chézine basin located in north-western France, in the suburbs of Nantes. It is being used to understand the impact of urban development on the topographic course of surface waters.

5.1 Introduction

The extension of urban areas is altering the hydrological behaviour of urban and peri-urban areas. These areas are located on the outskirts of cities, where coexists both urban and rural environments. Since the fifties, impermeable grounds have been spreading against farming and natural land. Coming with new urban planning, they interfere with the dynamics of runoff water flows. The course of the water gets diverted by new obstacles (roads, buildings, walls ...). The surface stream is poured into a watershed, creating new concentration and accumulation areas.

E. Bocher and M. Neteler (eds.), *Geospatial Free and Open Source Software in the 21st Century*, Lecture Notes in Geoinformation and Cartography,
DOI 10.1007/978-3-642-10595-1_5, © Springer-Verlag Berlin Heidelberg 2012

In order to understand the impact of these changes, many research studies have been using digital elevation models (DEM) (O'Callaghan and Mark 1984; Fair-field and Leymarie 1991; Costa-Cabral and Burges 1994; Jenson and Domingue 1988). These models give a representation of ground surface elevation, most often from a regular square grid (raster). They are used to build a drainage network, representing the surface water pathways in a watershed. They are then converted to one or more directional values for each pixel of the DEM. This constitutes the first step of distributed hydrological modelling. It defines the behaviour of water flows inside the watershed and provides means to compute additional variables like the distance to the river, the contributing and accumulation areas.

These methods to build a drainage network from a topography were used within a GIS (Geographic Information System) that uses external data (road locations, land use distribution), and adapted to take into account the disturbance produced by human activities (Mérot et al. 1999; Duke et al. 2003; Takken et al. 2001). Unfortunately they are based on a regular grid representation of the land and thus only provide partial answers, with limitations related to the modelling of complex areas (Vivoni et al. 2004; Tarboton 1997; Bocher 2005). This happens in urban areas because anthropogenic artefacts are particularly present and are very different and irregularly spread out (dense, intermittent, industrial or vegetation-based areas ...). Indeed, a DEM based on a regular square grid has some limits:

- The computation of the flow direction is limited by the neighbouring pixels: the directions can only vary by steps of 45°.
- The spatial representation of the objects is approximated: the whole pixel has the same elevation.

The external data has to be converted to a raster structure with the same resolution than the DEM. It is then difficult to maintain coherence between the original geometric object and its numerical conversion unless it is at a very high spatial resolution.

In urban and peri-urban areas, the use of a regular DEM is a strong limitation with respect to the number of elements that affect the water pathways. Moreover, it does not permit to take advantage of the Urban Data Banks (UDB) accuracies that describe the city with a high spatial resolution where features are described using vector geometries (Rodriguez et al. 2003).

In order to break free of these constraints and improve the modelling of runoff water pathways in urban areas, several authors suggested using a Triangular Irregular Network (TIN) (Tukker et al. 2001; Vivoni et al. 2004; Bocher 2005). A TIN is an irregular grid made up of three primitive types: nodes, edges and triangles, interconnected in topological relationships (Laurini and Thomson 1992). Every edge contains a reference to its start and end nodes and to the two triangles it is a part of.

A TIN has several benefits for the study of runoff water pathways:

- The geometric structure can be adapted to the surface relief. The amount of triangles can be increased in break slope areas, or reduced in more regular ones.
- It is thus freed from the constrained raster structure and its regular square grid limitations.
- Using the topological links, it is possible to make logical computations and quickly deduce morphological properties: thalwegs, ridges, flat triangles at the bottom of valleys …
- Surface topography is always qualified with a coherent planar element: a triangle. A single vector indicates the flow direction: the steepest topographic gradient.

This paper focuses on run-off water pathways spatial modelling. It intends to provide a better description taking into account natural and anthropogenic artefacts that modify the run-off pathways at the watershed scale. Due to the grid-based limitations, we propose an adapted version of the TAnaTo model for urban areas, named TAnaTo2.

5.2 TAnaTo Overview

TAnaTo is a TIN-based (Triangular Irregular Network) approach to evaluating the impact of agricultural practices and human-made features such as hedgerows, roads and land cover parcels in rural areas. It has been developed by Bocher (2005). The singularity of the approach lies in the definition of a new graph of cells called basins graph that expresses the relationships between linear constraints and topography. It represents topographic and anthropic flow directions. This basins graph is used to highlight the sequences of processes and the system networks that structure the organization of the water pathway.

The basin graph is based on a TIN with three types of constraints (Table 5.1). The constraints are vector geometries (points, lines and polygons) described with X, Y and Z coordinates.

Table 5.1. Description of the objects in the TIN

Feature (added in the TIN)	Geometry	Description
Obstacle	Line, polyline	Represents obstacles above the ground like a wall or a bank. A height parameter describes the height above the ground
Natural water network	Line, polyline	Represents rivers and streams
Anthropogenic water network	Line, polyline	Only represents open-air gaps
Parcel	Polygons	Represents land usage. Streaming ratios for winter and for summer are given for each parcel
Road	Line, polyline	Represents road sides. Each side contains two boolean attributes to detail if the stream follows the side and if it can overflow onto other geometries of the triangulation

The TIN is based on a Delaunay triangulation. The method produces three files containing nodes, arcs and triangles. Arcs and triangles features contain their objects attributes as well. It is thus possible to apply a filter to the list of arcs to get all arcs that represent obstacles. An example of constrained triangulation is given on Fig. 5.1.

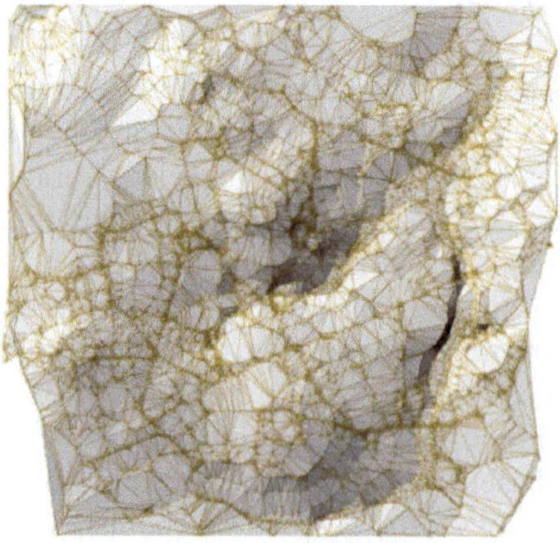

Fig. 5.1. Constrained TIN. Constraints are highlighted in *brown*

The arcs are then scanned and morphologically sorted by the steepest slope of their adjacent triangles. This step connects all geometries of the TIN to one another and provides a field topographic graph (Fig. 5.2). The basin graph can then be deducted from this graph using the attributes used when triangulating (object types, heights, etc.). It consists in locally changing the topographic connections depending on the type of the arc. Thus if the slope gradient points to an obstacle arc, this will be changed to follow the obstacle in a downhill direction. On Fig. 5.3, orange arcs represent the diversion of the water caused by the presence of a hedgerow.

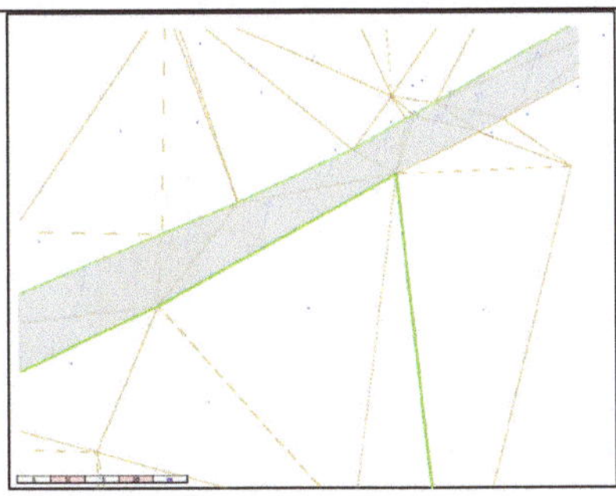

Constrained TIN
Green = hedgerow, gray = road, arrow = slope direction

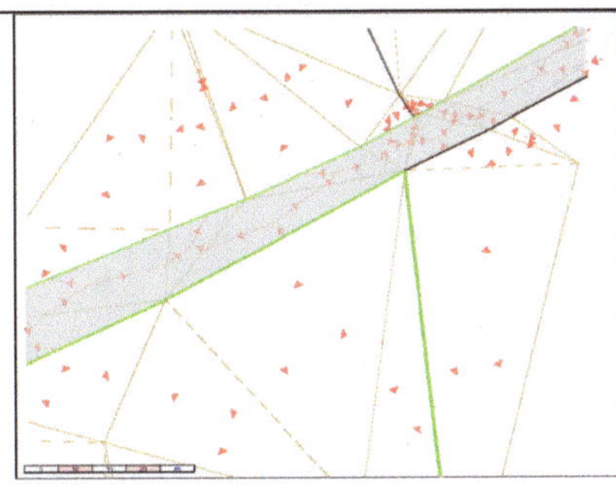

Topographic graph
Red arrow = geometries TIN connection

Fig. 5.2. Topographic graph

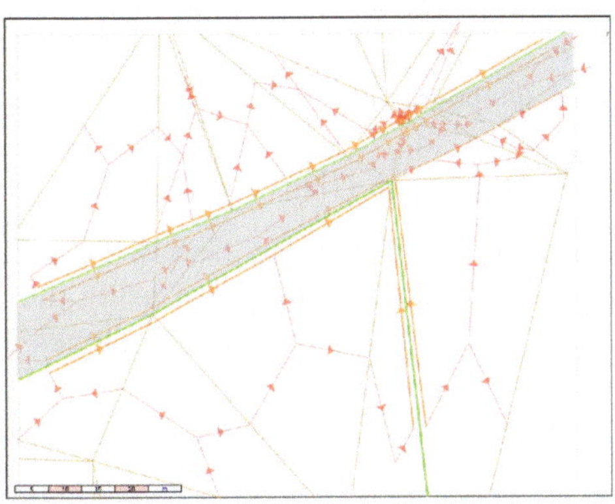

Fig. 5.3. Basin graph

Both topographic and constrained basin graph are exploited to compute spatial and statistical indicators like closed watershed and artificial runoff accumulation point. With TAnaTo the user can compare on the same river outlet a topographic watershed and a constrained watershed. A constrained watershed take into account the topographic direction modified by anthropogenic artefacts. It is built on top of the basin graph. Figure 5.4 shows an example of both a topographic basin graph and a graph constrained with hedgerows. The shape and size of the basin graph changes because some hedgerows prevent water from joining the river.

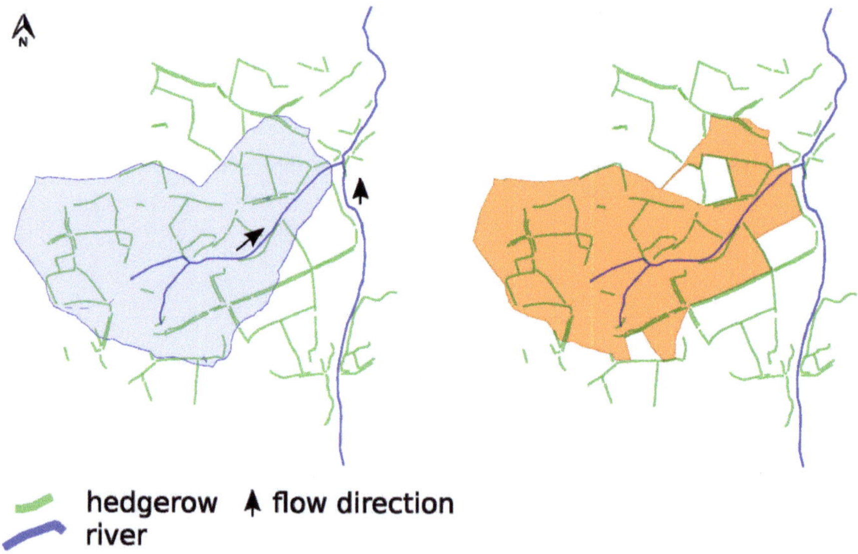

Fig. 5.4. On the *left side* a topographic basin on the *right side* a constrained basin

TAnaTo takes advantage of the TIN representation:

- Unambiguous derivation of the water flow from each triangle. Flow paths over a TIN surface can be computed using the steepest lines of descent of the TIN faces in any orientation (Jones and Grant 1996; Petzold et al. 2001).
- Accuracy preservation of landform elements. They are embedded as break lines into the TIN.

Nevertheless, because of input features (roads, ditches, parcels and hedgerows) and spatial dimension modelling constraints – TAnaTo supports only ground surface features – some improvements are still needed. Indeed, in urbanized areas, the main contribution to the catchment response stems from water flows over impervious surfaces and in undergrounded storm drains (Rodriguez et al. 2003). The former, impervious surfaces, is already described in TAnaTo using polygon parcels and runoff coefficients. It traduces the physical soil response. But the second one is not taken into account because it is a typical urban element and as said before TAnaTo was created with rural areas in mind.

Therefore, the structure of the TAnaTo model and the basin graph computing methods were adapted to:

- Take into account underground elements like sewer systems that collect and deliver water in urban areas.
- Take into account crossings (bridges, pipes) that can move water flows from one slope to another, which changes the drainage basin and as a consequence the contributing area.

The next section describes the improvements added to the model. These changes led to the TAnaTo2 version.

5.3 The TAnaTo2 Model

5.3.1 Data Input

To convey the impact of urban human artefacts, new geographic data has to be added to the model.

We can distinguish between two data types:

- Elevation data
- Topographic data

Elevation data represent the elevation distribution. It is represented with lines and points in the model.

Topographic data are of two types: linear data – for networks – and surface data – for land cover classification (Fig. 5.5).

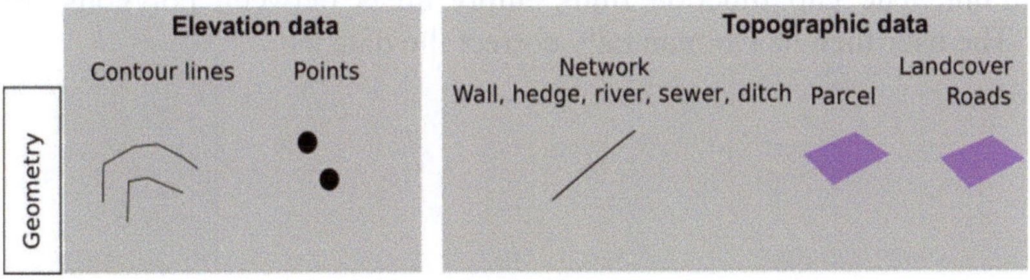

Fig. 5.5. Data input

As in the old TAnaTo model, linear data represents two kinds of geographic features (obstacles and both artificial and natural drains), as well as the sewer network. The sewer network is represented as a set of oriented polylines with entry points at ground level and exit points connected to artificial of natural drains. The rest of the sewer network is located underground.

In order to be integrated into the model and triangulated by the mesh builder of TAnaTo2, geographic data has to adhere to some topological constraints, both intern and extern. Table 5.2 gives a list of three constraints to be respected by networks: equal, overlap and self-intersection.

Table 5.2. Topological rules for networks (F = False, T = True)

	Natural network	Artificial network	Sewer network	Obstacle
Natural network	FFF	FFF	FTT	FFF
Artificial network	FFF	FFF	FTT	FFF
Sewer network	FTT	FTT	FFF	FTT
Obstacle	FFF	FFF	FTT	FFF

Table 5.2 gives the rules for the intersection of a line and a column entry. For example, geometries of the “natural network” layer (FFF):

- Cannot be duplicated (F for equal).
- Cannot overlap (F for overlap).
- Cannot intersect without a node (F for self-intersection).

Figure 5.6 gives graphical examples of overlap and self-intersection constraints.

For land cover data, the same analysis with topological constraints is done except the self-intersection constraint is replaced with the gap constraint. The gap function finds empty areas between polygons (Fig. 5.7). The user then has to manually correct the data.

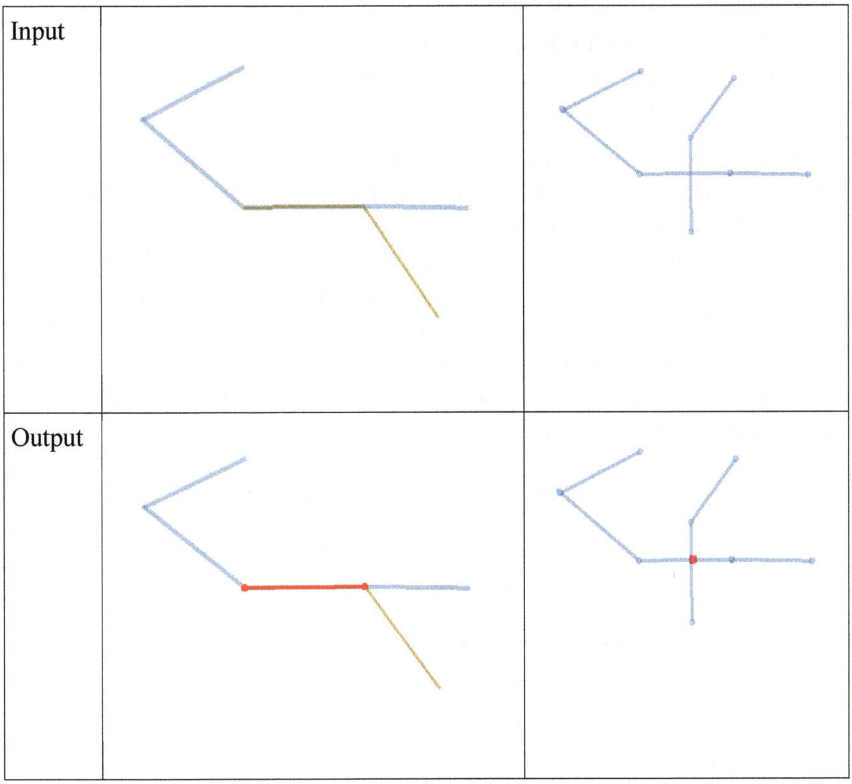

Fig. 5.6. Data network quality constraints

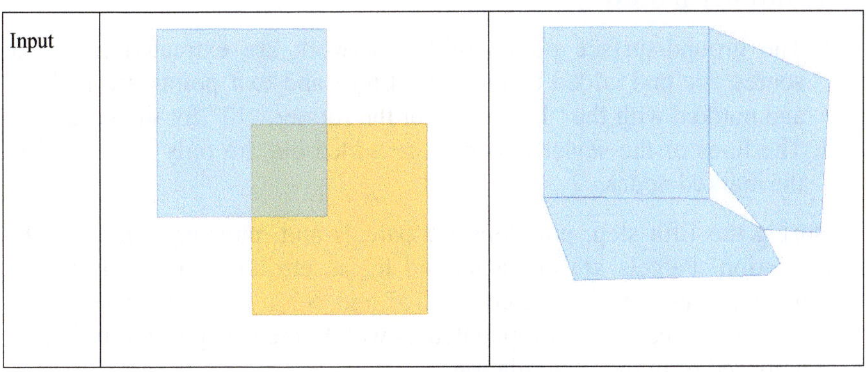

Fig. 5.7. Land cover data quality constraints

5.3.2 TIN Building

The TIN is built in five steps:

1. A TIN is created with altitude attributes.

2. Flat areas are removed.
3. Surface networks are incorporated.
4. Sewer entry and exit points are inserted.
5. Land use data is added.

The first three steps already were implemented in the previous version of TAnaTo, they will not be described here. The reader can look into Bocher (2005) for more information. The result of these three steps is a TIN that contains altitudes and surface networks, i.e. obstacles and both artificial and natural drains. So as to facilitate comprehension, this TIN will be called TINTopo_Network from now on. Networks are integrated into it as constraints. An integer value is attributed to every constrained arc of the TIN to characterize the object type it represents (Table 5.3). This value is stored in a field called "type".

Table 5.3. Type value for each type of constraint

Category	Data	Type value
Elevation	Contour lines	3
Topographic	Obstacle	5
	Natural drain	7
	Artificial drain	9

The forth step is about adding to TINTopo_Network the sewer network. It is a two-step process:

1. The ground-surface points of the network are extracted from the source file and added to the TIN. Entry and exit points are isolated and marked with the "11" value for the former, "13" for the latter.
2. The lines of the sewer network are added but are only connected to the marked nodes.

During the fifth step, polygons for parcels and roads are added to the triangulation. Parcels are characterized by an attribute, either "rural" or "urban", with respectively the codes "15" and "17". The "19" code is used for roads. These codes are propagated as well to the triangles the polygons are composed of in the triangulation.

Figure 5.8 shows an example of a TIN produced with contour lines (dotted red lines), obstacles (green lines), a natural drain network (blue lines) and parcels (grey polygons).

Fig. 5.8. Full data triangulation

The TIN features are stored in three files: “nodes”, “edges” and “triangles”. These files are bound to one another using foreign keys. The “edges” file is the centre of the TIN. It contains references to IDs of the nodes and IDs of the triangles. It also stores the IDs of the start and end points of the edge, and IDs of the neighbouring triangles.

5.3.3 Graph Computing

Just as with the previous TAnaTo version, two graphs are produced: a topographic graph and a basin graph. The edges are scanned so as to build the topographic graph. For every edge, the algorithm finds, for both neighbouring triangles, the direction of the steepest slope gradient. If both aim at the edge, then the arc is a thalweg. On the other hand, if none of them do, it is a ridge. Altogether, 12 morphological configurations were identified (Bocher, 2005).

The basin graph is built by altering local water flow directions depending on the surface objects they encounter. To change the directions, the object type of the edge is considered (Sect. 5.3.2).

Both graphs are stored in files on disk (Sect. 5.4.2). A graph is represented as a set of nodes referencing the TIN’s features. Nodes are HydroCell objects:

- Ncell for nodes
- Ecell for edges
- TCell for triangles

Every Hydrocell stores in two separated lists its upstream and downstream Hydro-cells. It also stores the “type” code of the underlying object (for example “5” for an obstacle). Some Hydrocell types have additional methods. For example, an Ecell or Tcell object has a getSlope() method to get its slope in degrees.

A schema of the two graph produced from a simple TIN is shown on Fig. 5.9. It can be noticed that triangle B is disconnected from the triangle C. The water from B has been rerouted because there is a wall between B and C and descending from NCell 3 towards the NCell 2.

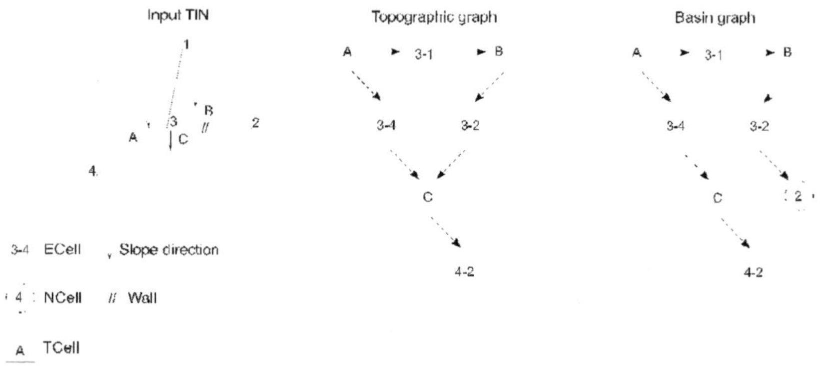

Fig. 5.9. Schema of the graphs

Using the usual graph browsing techniques, the basin graph is used to:

- Find outlet cells (hydrocells with no downstream cells)
- Find the contributing area of a cell (watershed)
- Extract the path of the water flow from a cell

5.4 TAnaTo2 as Open-Source GIS Plugin

TANATO2 is a GPL3-licensed application plugged on the OrbisGIS open-source GIS software. One of the motivations for the use of open-source GIS software was:

- To contribute to a common platform with the intend to federate scientific research in the field of geomatics
- To facilitate the dissemination of scientific models and knowledge without cost and licence restrictions

- To take profit of features on which other researchers are working (such as spatial SQL language, space-time modelling or multiscale data integration ...) which help spatial analysis processing in TAnaTo2

5.4.1 OrbisGIS

OrbisGIS[1] is a Geographic Information System dedicated to scientific modelling and experimentation. It has been developed by the IRSTV (Institute for Research on Urban Sciences and Techniques, FR CNRS 2488) since 2007 within a local PRF (Federative Research Project) on urban modelling which aims at providing methods and tools to face urban issues.

OrbisGIS is based on open-source libraries. It is under the GPL3 license. It aims at pooling methods and processes in use within IRSTV's research teams, whatever field they are in (sociology, civil engineering, urban architecture, geography, economics, environmental sciences ...)

In addition to the usual features of a GIS (map viewing, cartography, data edition, database access), OrbisGIS's special feature lies in its GDMS library (Generic Datasource Management System) (Bocher et al. 2008). GDMS is an abstraction layer that can read and write both geographic and alphanumeric data in several different formats (Shapefile, CSV, PostGIS database...). Data is accessed through an API (Application Programming Interface) that uses a java object named "DataSource". Apart from the API, the "DataSource" objects can be accessed using an internal SQL language. Following the example of relational database systems, this language can be improved by adding custom functions. It contains all geometric operators and predicated listed in the "Simple Features SQL" specification (Herring 2006). This brings GDMS to the level of the spatial addition PostGIS. See Bocher et al. 2008 for more detailed information about GDMS.

5.4.2 TAnaTo2 Tools

TAnaTo2 has been developed so that its processing algorithms are available as SQL functions for GDMS: building a TIN, building the graphs and using them. These functions are manipulated in the SQL console of the OrbisGIS software. Figure 5.4 shows an SQL instruction to produce a constrained TIN from contour lines. After this function has been executed, the three data sources that were produced are listed in the Geocatalog,

---

1 http://www.orbisgis.org

OrbisGIS's data catalogue. The user can then drag-and-drop them from the Geocatalog to the TOC (Table Of Contents) of a map window (Fig. 5.10) in order to display them.

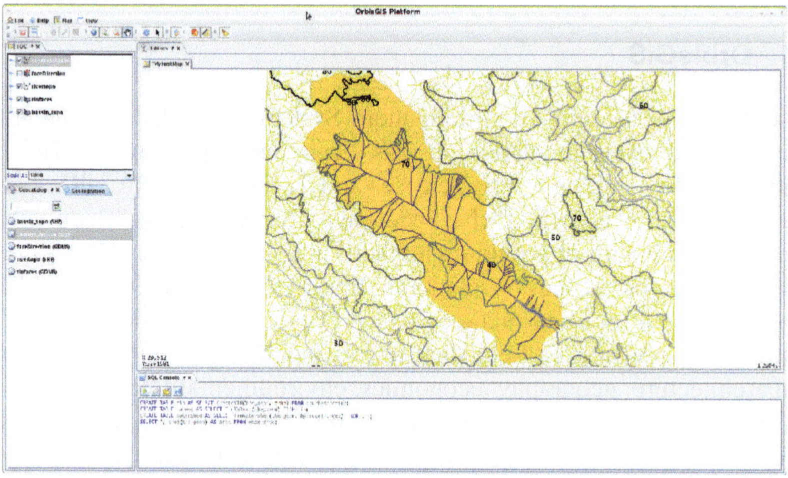

Fig. 5.10. Using Tanato2 with OrbisGIS's SQL console

Table 5.4 lists the SQL functions used to build the TIN.

Table 5.4. Functions used to build TINs

Function	Description	Usage
ST_DT(GEOMETRY [,DOUBLE])	Builds a Delaunay triangulation from the geometries in a table. An optional double value can be used to constraint the triangles' angles	SELECT ST_DT(the_geom, 30) FROM myLayer
ST_CDT(GEOMETRY[,BOOLEAN, DOUBLE])	Builds a constrained Delaunay triangulation from the geometries in a table. An optional boolean value is used to mean if flat triangles have to be deleted. An optional double value can be used to constraint the triangles' angles	SELECT ST_CDT(the_geom, true) FROM myConstraintLineLayer

Once the TIN is built, the user can the run the functions "ST_HydroTopoGraph" and "ST_HydroBasinGraph" to compute the topological and basin graphs. These two functions produce four new data sources ("graph" tables) that store the relations between the TIN's elements (Sect. 5.3.3). No specific file format is used to save the graphs. TAnaTo2 uses two tables per graph: one for nodes, the other one for connections between them.

The "ST_HydroOutlet" function is then used to extract exit points of the graph to a table of points. The function is applied on the graph's "arc" table. Displaying the exit points in OrbisGIS can be used to visually find the elements disrupting the water flow. To get into more details, the user can display the steepest slope directions of the arcs and triangles using the function "ST_HydroSlopeDirection".

Finally, the "ST_HydroWatershed" function can be used to analyse the global effect of city planning elements on water flows. Indeed this function computes the watershed of a particular node of the graph by collecting every cell of the TIN that is upstream of the exit node. This function is essential to understand the impact of urban planning on the actual area of watersheds.

Table 5.5 sums up the SQL functions detailed above.

Table 5.5. Graph analysis functions

Function	Description	Usage
ST_HydroOutlet(GEOMETRY)	Extract all outlets based on an input graph edge table	SELECT ST_HydroOutlet(the_geom) from topoEdgeGraphTable
ST_HydroSlopeDirection (GEOMETRY)	Extract all gradient slope direction as vector features	SELECT ST_HydroSlopeDirection (the_geom) from topoEdgeGraphTable
ST_HydroWatershed(INTEGER)	Extract a watershed based on a graph node id or a point geometry	SELECT ST_HydroWateshed(12) FROM topoGraphLayer

5.5 Case Study

5.5.1 Study Area and Data

To illustrate this research, TAnaTo2 was applied on a peri-urban basin called Chézine, located in north-western France (suburbs of Nantes). The Chézine area is a 29-km² basin with a 15 km-long river (Fig. 5.11). Nevertheless, because of unavailable data, the study area has been limited to a smaller watershed that covers approximately 500 ha (Fig. 5.12).

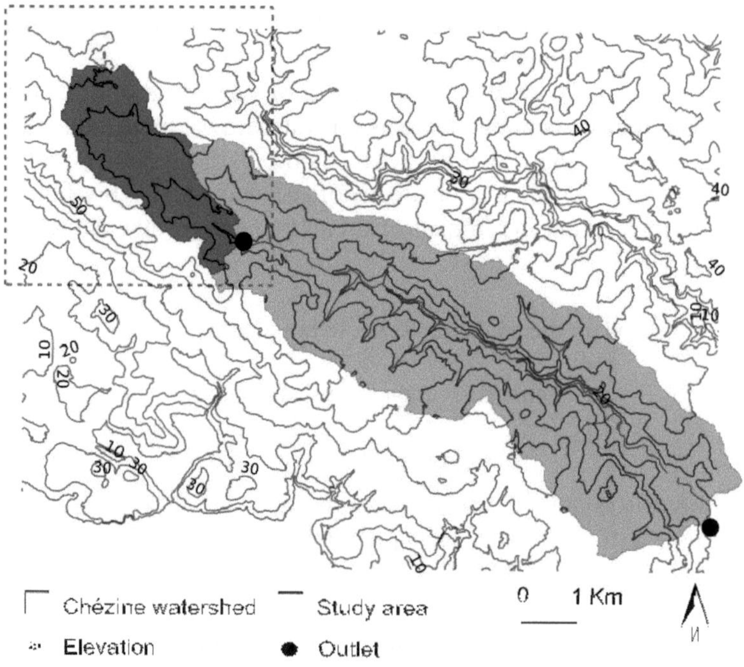

Fig. 5.11. Study area location

Table 5.6 describes the data used for this example. In addition to the attributes given in the table, each data source contains an additional “type” attribute. It is used during the triangulation. The data are stored in the ESRI Shapefile file format.

Table 5.6. Characteristics of the data used

Data	Description	Features statistics
DEM	Digital Elevation Model from the French geographic institute (IGN)	Pixel size : 25 m Max Z : 86 m Min Z : 7 m Mean Z : 61.4 m
Contour lines	Topography is represented by a set of polylines. Altitudes are stored in an “altitude” field, and in the geometries. The lines were extracted from the DEM every 10 m	Number : 16 Max length : 23,521 m Min length : 47.5 m Mean length : 4,935 m
Roads	Road lines provided by IGN in the context of the reconstruction of the national database: RGE. The road lines are polylines with altitude data. A “largeur” column gives the width of the road	Number : 782 Max length : 3,000 m Min length : 1 m Mean length : 186 m
River network	Hydrographical network represented by polylines in 2.5D.	Number : 57 Max length : 764 m Min length : 7 m Mean length : 111.9 m
Ditch network	Ditch network, represented by a polylines in 2.5D. The network was collected on site.	Number : 966 Max length : 1,179.8 m Min length : 6.3 m Mean length : 97.5 m

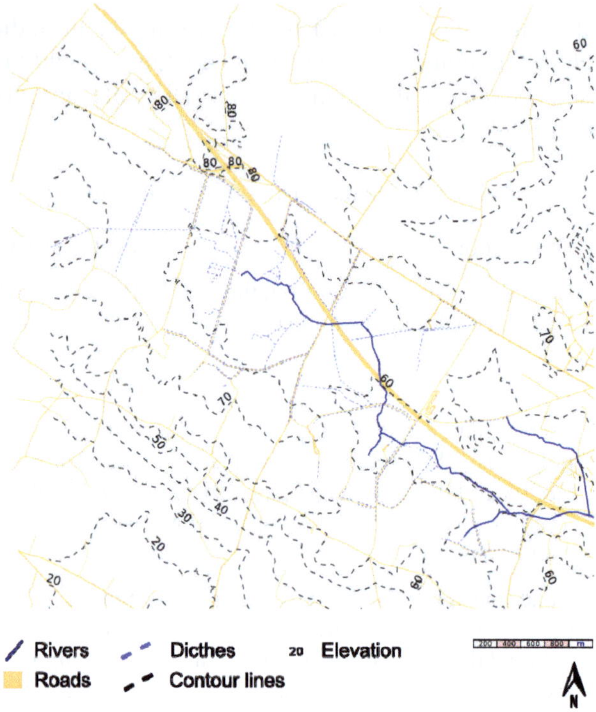

Fig. 5.12. Map of the basin in use

5.5.2 Scenarii Model

In order to study the way TAnaTo2 evaluates the impact of anthropogenic elements, five modelling schemes were applied (Table 5.7).

Table 5.7. Model schemes and data being used

Model	Data input
1	DEM
2	Contour lines
3	Contour lines + river network
4	Contour lines + river network + ditch network
5	Contour lines + river network + ditch network + roads

This is done to evaluate the contribution of every element – river, ditches and roads – in the behaviour of surface water flows within the watershed. In order to do this, we compute for every model the contributing area, from the same exit point of the basin. The coordinates of the exit point is x = 345,461.321 and y = 6,694,459.291.

5.5.3 Results

The first model is computed using a set of raster hydrological processes available in OrbisGIS (LeDuc et al. 2009). The sequence is described in Table 5.8. It's divided in three SQL statements. The watershed is based on the D8 flow direction method defined by O'Callaghan and Mark 1984.

Table 5.8. Computing the watershed with model 1

```
– The raster DEM is filled to delete sink and flat areas using Planchon and Darboux 2002
CREATE TABLE filledDem AS SELECT ST_FilledDem(raster, 0.01) AS raster FROM dem;
– A raster flow direction is build based on the D8 method defined by O'Callaghan and D. Mark., 1984.
CREATE TABLE direction AS SELECT ST_D8Direction(raster) AS raster FROM filledDem;
– The raster that contains the watershed for the outlet point () is computed.
CREATE TABLE watershed AS SELECT D8Watersheds(raster, ST_GeomFromText('POINT (345461.321 6694459.291)')) AS rs FROM direction AS;
– The raster watershed is converted to a polygon. No data values are ignored.
CREATE TABLE dem_basin AS SELECT ST_RastertoPolygons(raster) FROM watershed;
```

The second model consists in triangulating the contour lines and removing flat triangles. This is done in three steps (Table 5.9).

Table 5.9. Computing the watershed with model 2

```
– Triangulation of the contour lines and removing flat triangles
SELECT ST_CDT(the_geom, true) FROM contourLines;
– Topological graph computation
CREATE TABLE contourLines_topoGraph AS SELECT ST_HydroTopoGraph() FROM contourLines_nodes, contourLines_edges, contourLines_triangles;
– Watershed computation from a given point
SELECT ST_HydroWatershed(ST_GeomFromText('POINT(345461.321 6694459.291)'))
FROM contourLines_topoGraph;
```

The third model is the same except the graph is built using the "river" constraint (type 7) (Table 5.10).

Table 5.10. Computing the watershed with model 3

```
– Merge contour lines and river tables
CREATE TABLE constraintData AS (SELECT the_geom, type FROM contourLines) UNION
(SELECT the_geom, type FROM river);
– Constrained delaunay triangulation
SELECT ST_CDT(the_geom, true) FROM constraintData;
– Basin graph computing
CREATE TABLE constraintData_basinGraph AS SELECT ST_HydroBasinGraph() FROM
constraintData_nodes, constraintData_edges, constraintData_triangles;
– Compute the watershed from an outlet
   SELECT ST_HydroWatershed(ST_GeomFromText('POINT(345461.321 6694459.291)'))
FROM constraintData_basinGraph;
```

Both models 4 and 5 use the same script as model 3. Only the initial data changes (constraintData). Model 4 adds rivers and the ditch network (types 7 and 9), and model 5 adds rivers, the ditch network and roads (types 7, 9 and 19).

A map of the produced watersheds is shown on Fig. 5.13.

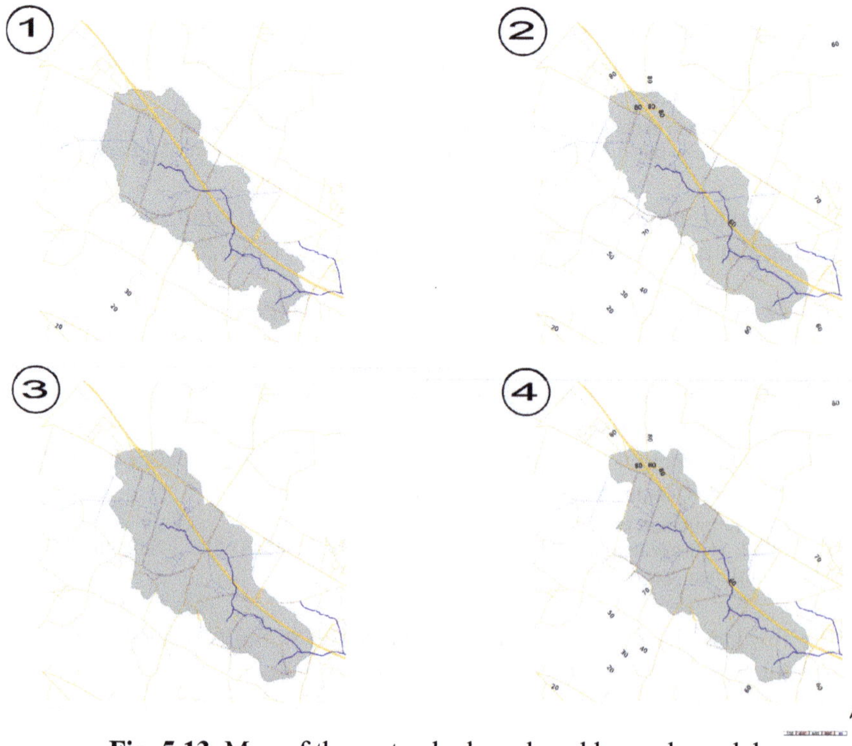

Fig. 5.13. Map of the watershed produced by each model

The analysis of the area of these watersheds shows that the choice of the method of watershed computation – D8 or TIN – has a large influence on the obtained surface connected to the river. The topographic basin produced by the TIN is smaller than the one produced by the D8 method. There is a 44 ha difference. Maps 1 and 2 (Fig. 5.13) show local similarities with finding contour lines, notably north of the main road, but differences in the southern parts: the basin is more contracted on basin 2.

About surface elements, models 3 and 4 show that these elements do change the final area, as well as drainage areas (Fig. 5.13).

Taking into account the ditch network makes the drainage area increase by 61.4 ha (model 2 vs. model 3). The cartographic comparison of the two models (2 and 3) highlights several additional contributions areas. North-west of the watershed, the ditch network behaves as a drain. However this is not always the case in model 4. Indeed, taking into account the roads leads to a separation of some areas from the river. In the west parts, the watershed's shape gets smaller. This is confirmed by computing the global area of the watershed: it decreases by 30.4 ha between model 3 and 4 (Fig. 5.14).

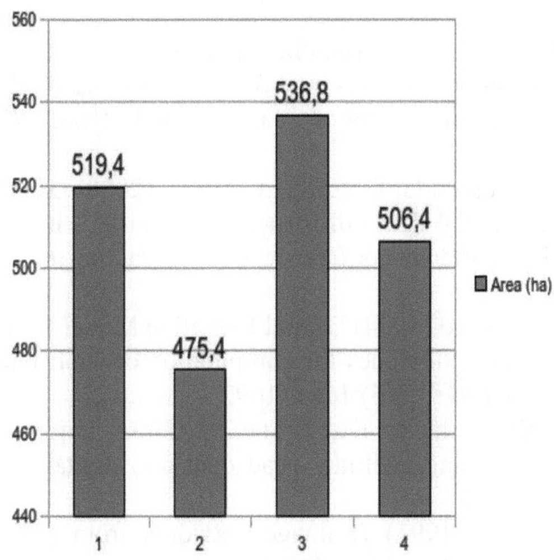

Fig. 5.14. Distribution of the watershed areas

5.5 Conclusion

This chapter introduced a new version of the TAnaTo model. TAnaTo2 spatially analyse the impact of anthropogenic urban planning on run-off water pathways using a TIN structure. The computation of four different watersheds confirms that linear elements have important in the control over surface water flows. The results corroborate the work of Mérot et al. (1999), Duke et al. (2003) as well as Bocher (2005), who showed the influence of the position and connections of networks on water flows. However these observations have to be confirmed by more detailed analysis that would enable us to view the water flows diverted by the obstacles or accumulation points and surfaces. These numeric results must be validated as well using actual on-site observations.

Acknowledgments

This research is as part of the AVuPUR project, funded by the French Agence Nationale de la Recherche (ANR) under contract ANR-07-VULN-01.

References

Bocher E (2005) Impacts des activités humaines sur le parcours des écoulements de surface dans un bassin versant bocager: essai de modélisation spatiale. Application au Bassin versant du Jaudy-Guindy-Bizien (22) Ph.D. thesis, Université de Rennes 2

Bocher E, Leduc T, Moreau G, González Cortés F (2008) GDMS: an abstraction layer to enhance spatial data infrastructures usability'. In: Proceedings of the 11th AGILE international conference on geographic information science – AGILE'2008, Girona

Costa-Cabral M, Burges S-J (1994) Digital Elevation Model Networks (DEMON): a model of flow over hillslopes for computation of contributing and dispersal areas. Water Resour Res 30(6):1681–1692

Duke GD, Kienzle SW, Johnson DL, Byrne JM (2003) Improving overland flow routing by incorporating ancillary road data into digital elevation models. J Spatial Hydrol 3:1–27

Fairfield J, Leymarie P (1991) Drainage networks from grid digital elevation models. Water Resour Res 27(5):709–717

Herring JR (2006) OpenGIS implementation specification for geographic information - simple feature access - Part 1: Common architecture, October 2006. www.opengeospatial.org/standards/sfs. Accessed 2 Sept 2009

Jenson SK, Domingue JO (1988) Extracting topographic structure from digital elevation data for geographic information system analysis. Photogramm Eng Rem S 54(11):1593–1600

Jones JA, Grant GE (1996) Peak flow responses to clear-cutting and roads in small and large basins, western Cascades, Oregon. Water Resour Res 32(4):959–974

Laurini R, Thompson D (1992) Fundamentals of spatial information systems. Academic Press, Toronto

Leduc T, Bocher E, González Cortés F, Moreau G (2009a). GDMS-R: Amixed SQL to manage raster and vector data. In GIS Ostrava 2009 - Symposium on Seamless Geoinformation Technologies, Ostrava, Czech Republic.

O'Callaghan J, Mark D (1984) The extraction of drainage networks from digital elevation data. Comput vision graph 28(3):323–344

Mérot Ph, Gascuel-Odoux C, Walter C, Zhang X, Molenat J (1999) Influence du réseau de haies des paysages bocagers sur le cheminement de l'eau de surface. Revue des sciences de l'eau 12(1):23–31

Petzold I, Gröger G, Plümer L (2001) Network planning using geomorphology. In: Proceedings of the 9th ACM (Association for Computing Machinery) international symposium on advances in geographic information systems. Atlanta, pp 167–172

Planchon O, Darboux F (2002) A fast, simple and versatile algorithm to fill the depressions of digital elevation models. CATENA 46(2–3):159–176

Rodriguez F, Andrieu H, Creutin JD (2003) Surface runoff in urban catchments: morphological identification of unit hydrographs from urban databanks. J Hydrol 283:146–168

Takken I, Govers G, Steegen A, Nachtergaele J, Guérif J (2001) The prediction of runoff flow directions on tilled fields. J Hydrol 248:1–13

Tarboon D (1997) A new method for the determination of flow directions and contributing areas in grid digital elevation models. Water Resour Res 33(2):309–319

Tucker G, Lancaster S, Gasparini N, Bras R, Rybarczyk S (2001) An object oriented framework for distributed hydrologic and geomorphic modeling using trian-gulated irregular networks. Comput Geosci 27:959–973

Vivoni E-R, Ivanov V-Y, Bras RL, Entekhabi D (2004) Generation of triangulated irregular networks based on hydrological similarity. J Hydrologic Eng 9(4):288–302

Part 2: Remote Sensing

Chapter 6 The E-Foto Project and the Research to Implement a GNU/GPL Open source Educational Digital Photogrammetric Workstation

Guilherme Lucio Abelha Mota, Jorge Luís Nunes e Silva Brito, João Araújo Ribeiro, Orlando Bernardo Filho, Marcelo Teixeira Silveira, Rafael Alves de Aguiar, Irving da Silva Badolato, Sarina Lustosa da Costa, Patrícia Farias Reolon

Master's Program in Geomatics, The Rio de Janeiro State University,
524 São Francisco Xavier Street, Room 5028-D, Maracanã,
Rio de Janeiro RJ, 20550-900, Brazil
efoto@yahoo.groups.com

Abstract. Photogrammetry is the science devoted to the geometric reconstruction of object space, based on photographic images. Among its techniques, stereophotogrammetry, which employs a pair of images to precisely estimate 3D coordinates in the object space, should be highlighted. The state of the art of photogrammetric equipment is represented by digital photogrammetric workstations (DPW), that are primarily hardware and software dedicated to photogrammetric purposes. The present chapter is devoted to the E-Foto project, which, to the best of our knowledge and belief, implements the only open source GNU/GPL digital photogrammetric workstation presently available. This project is originally concerned on education and is underlain on the synergy between development, teaching and research devoted to this field of science. It encompasses a team of undergraduate and Master students as well as Professors of The Rio de Janeiro State University. Besides the DPW, the E-Foto project products are: the only digital photogrammetry book edited in Brazil, the project's home page, the software usage tutorials, master dissertations, graduation projects and research papers.[1]

[1] http://www.efoto.eng.uerj.br

E. Bocher and M. Neteler (eds.), *Geospatial Free and Open Source Software in the 21st Century*, Lecture Notes in Geoinformation and Cartography,
DOI 10.1007/978-3-642-10595-1_6, © Springer-Verlag Berlin Heidelberg 2012

6.1 Introduction

Proprietary application software keeps source code as a secret, what may make it unsuitable to academic purposes. As a matter of fact, such applicative programs are like "black-boxes" that are merely limited to the graphical user interface (GUI) handling. This fetters their application to education, since users remain unaware of what occurs in the background. In addition, it is not possible to extend or customize software without source code access. Therefore, concerning research activities, the usage of proprietary software is also quite limited.

A common proposal for overcoming such limitations is the development and usage of open source software, inasmuch as free access to source code extends research and educational software capabilities. Although many branches of geospatial technology have been benefited by open source software, digital photogrammetry is a field that remains practically unexplored. An exception is E-Foto, project which, to the best of our knowledge and belief, develops the only open source environment for research, teaching and learning of photogrammetry that encompasses a Digital Photogrammetric Workstation (DPW).

In addition to costless DPW supplying for operators training, E-Foto motivates the photogrammetric algorithms examination, modification and inclusion of new solutions, as well. To summarize, by producing and sharing without limitations with its community software and documents, this project is bringing about the basis of a commons dedicated to digital photogrammetry.

The present chapter is devoted to give to the open source geospatial research community a presentation on the main achievements, impacts and expectancies of E-Foto. The next section brings an introduction about digital photogrammetry and DPWs, while the following one describes E-Foto DPW's development. The succeeding section is dedicated to the distinct software development subprojects. The section that follows presents the additional photogrammetric commons products and, on end, comes the chapter conclusion.

6.2 Digital Photogrammetry

This section intends to bring a brief description of Digital Photogrammetry, readers interested on additional information can consult Kraus (2000), Schenk (1999), Mikhail et al. (2001) and ASPRS (1997).

Photogrammetry concerns the reconstruction of the 3D space (object-space) based on 2D photographic images (image-space). In the 1990s, its state of the art had turned to digital photogrammetry, which aims at modeling and reconstructing the object-space through digitally acquired or digitized-frame images underlain by digital or computational methods. The equipment able to perform these tasks, the so-called Digital Photogrammetric Workstation, is a dedicated system for photogrammetric purposes (Heipke 2001).

Digital photogrammetry techniques are useful for different knowledge fields, from virtual reality up to medicine, besides the non-invasive modeling of archeological sites (Kraus 2000). Additionally, on the cartographic engineering field, it is employed for the recovering of the shape of both earth surface and land cover features, what is meant by photogrammetric mapping.

For simplicity reasons, the following presentation is based on homogeneous coordinates of points, vectors and projective spaces. Two geometric transformations applied on cascade underlay the photogrammetric mapping process. Together, these transformations transport digital image-space coordinates to object-space coordinates. Both are affine transformations, the former transformation $A_{io}(\mathbf{x}) = \mathbf{M}^{io}.\mathbf{x}$ converts digital image 2D coordinates to the reference space of the sensor, where $\mathbf{x} = [col, row, 0, 1]^T$, while $\mathbf{M}^{io}$ is the 4 × 4 matrix that express such transformation. A_{io} leads to $\mathbf{x}' = [\xi, \eta, -f, 1]^T$, sensor space coordinates, where ξ and η are the 2D coordinates expressed on the sensor projection plane, while f is the sensor focal length. This outcome is submitted to the latter transformation, $A_{eo}(\mathbf{x}') = \mathbf{M}^{eo}.\mathbf{x}'$, leading to $\mathbf{x}'' = [x, y, z, 1]^T$, where $\mathbf{M}^{eo}$ is also a 4 × 4 matrix. In digital photogrammetry, the former and the latter transformations are respectively known as interior and exterior orientations.

Matrices $\mathbf{M}^{io}$ and $\mathbf{M}^{eo}$ are defined by parametric models, respectively by the columns matrices $\mathbf{X}^{io} = [a_o, a_1, a_2, b_o, b_1, b_2]^T$ and $\mathbf{X}^{eo} = [\varphi, \omega, \kappa, x_0, y_0, z_0]^T$, where a_i and b_i , for $i \in [0, 2]$, are polynomials coefficients, φ, ω, κ are the Euler angles that express the attitude of the sensor and x_0, y_0, z_0 represent the position of the camera's perspective center. All these parameters are computed in the object space reference system. The estimation applies the least squares principle, which determines the $\mathbf{X}$ matrix that best fits to the sample data. It must be mentioned that the estimation of $\mathbf{X}^{io}$ employs the linear least squares, while $\mathbf{X}^{eo}$ demands the usage the non-linear least squares approach.

This cascade model leads, as a function of the pixel coordinates, to the equation of the line that connects the corresponding object position to the sensor center, the so-called collinearity principle. Moreover, the spatial intersection of two lines, relative to the same object-space point and a pair of images acquired with stereoscopic parallax,[2] turns possible the precise estimation of the object-space coordinates of such point. Fig. 6.1 presents the usage of spatial intersection for point ***B*** coordinates determination. Indeed, such approach is employed in several photogrammetric operations, for instance, stereo-plotting (extraction of cartographic features) and Digital Terrain Model (DTM) extraction.

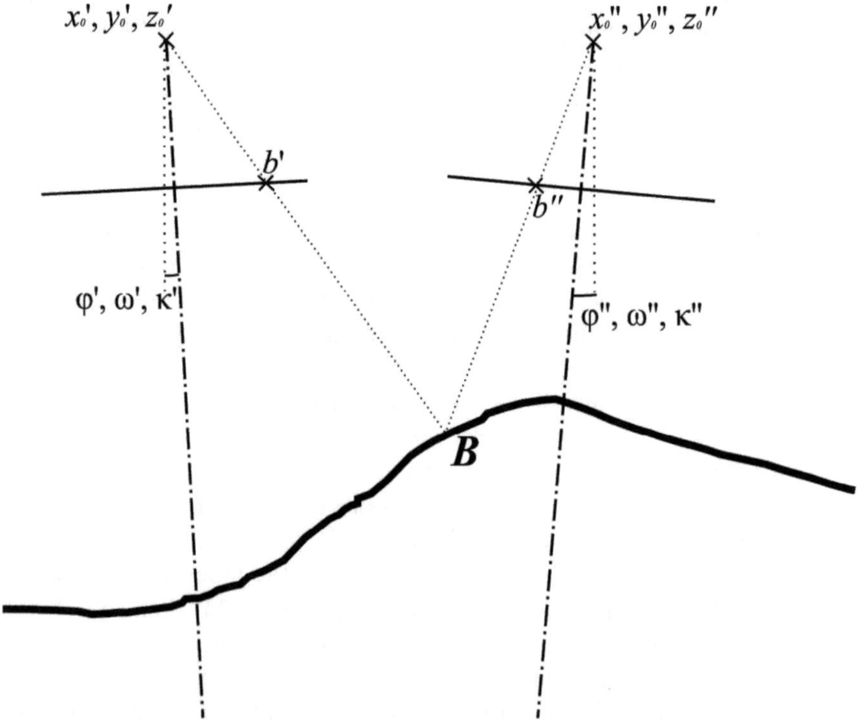

Fig. 6.1. Spatial intersection to the determination of the object-space coordinates of point B

On the other hand, the transformation that transport object-space coordinates onto image-space is given by $[A^{io}]^{-1}(P(\mathbf{x}''))$, where $[A^{io}]^{-1}$ is the inverse interior orientation transformation, while P is a projective transformation that brings coordinates from object-space to the sensor system. The elements of matrix $\mathbf{M}^{\mathbf{p}}$, which defines the P transformation, can be derived from $\mathbf{X}^{\mathbf{eo}}$ together with sensor parameters. That approach

[2] Parallax, in photogrammetry, can be defined as the apparent relative displacement of objects caused by changing the point of view.

is employed during the orthorectification (generation of orthogonal perspective images) and stereopair normalization, that delivers a stereopair registered in epipolar geometry (Schenk 1999).

6.2.1 Digital Photogrammetric Workstations

Photogrammetry has three main components: (1) photographic and ground control data; (2) photogrammetric techniques, algorithms and software; as well as, (3) hardware. Photogrammetric hardware has evolved along the time, and can be divided into three main architectures or technologies (Kraus 2000): (1) analog plotters, which were used in the first-half of the twentieth century. Those devices were mainly composed by mechanical parts and optics (prisms, mirrors, and lenses); (2) analytical plotters, which original solution was proposed in the early 1950s by Uki Helava, and; (3) Digital Photogrammetric Workstations, whose first solution became available in the late 1980s.

DPWs are computer workstations specially designed for performing photogrammetric tasks, operations and calculations. Their main hardware components are: (1) special video displays, either CRT or LCD; (2) stereoscopic viewing devices; (3) graphic processors and cardboards; in addition to, (4) I/O devices comprising keyboards, tablets, scanners, plotters, and special 3D mouses. Those who need further details on the components and architecture of DPWs can refer to ASPRS (1997) and Heipke (2001).

Although the hardware components play a very important role in the architecture of the DPWs, the core of them is, undoubtedly, the photogrammetric software available at each DPW solution. The photogrammetric software components make the biggest difference among the commercial solutions. This characteristic is due to the level of automation of photogrammetric tasks available in each solution and is, unavoidably, linked to the prices of those solutions. One could state that commercial DPW solutions (hardware, software, maintenance and training) range from U$ 20,000.00 up to U$ 120,000.00, depending upon the number of software licenses, hardware configuration and architecture. This range of DPW's prices has prohibited – or at least has inhibited – academy to purchase, for research and educational purposes, full capability DPW solutions in many countries. We think that this sentence is true independently of the policies of photogrammetric software companies toward the offering of special discounts and prices for educational configurations of photogrammetric workstations. Additionally, to our knowledge, educational licenses of photogrammetry software, when purchased, generally could not support the addition and integration of photogrammetric functionalities. For instance, many DPWs solutions – with a few exceptions– cannot be integrated to Geographic Information Systems (GIS).

6.3 E-Foto DPW Development

The E-Foto project started in 2002, with an undergraduate student in cartographic engineering, in The Military Institute of Engineering, in the City of Rio de Janeiro, in Brazil. The main goal of that work was designing and implementing an initial version of a DPW for educational purposes. That was a milestone of E-Foto project, it started to break of the paradigm of using proprietary photogrammetric software. The E-Foto software package is developed in an open source approach and follows the General Public Licence (GNU/GPL) principles established by the Free-Software Foundation: liberty to use for any purpose, to copy, to modify, and to distribute. This approach was sought to be useful to give free access to the source code to those who are interested in using, learning or even developing digital photogrammetry software solutions.

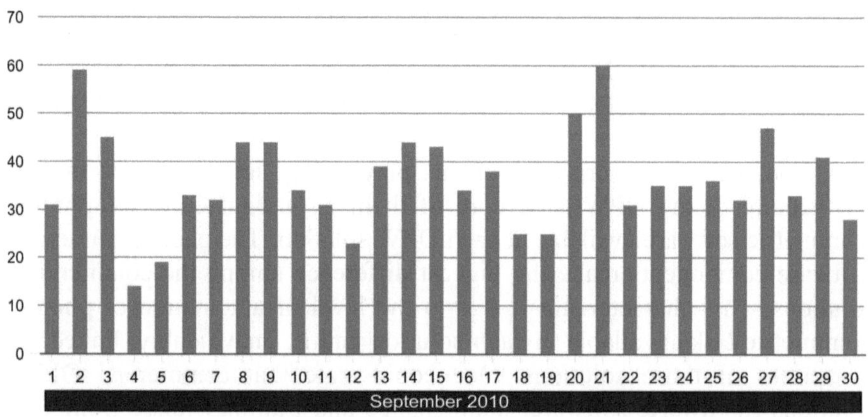

Fig. 6.2. Daily statistics about the E-Foto Homepage visits in September 2010

In 2003, the E-Foto project shifted to The Rio de Janeiro State University, when it earned a financial support of U$50,000.00 by the Brazilian Research Council, the CNPq. That funding granted by the E-Foto project permitted to set up the necessary infrastructure for software development. Besides this funding, the E-Foto project is also being financially supported by both Brazilian Federal Government and by The Rio de Janeiro State Government. The average funding resources earned along the past 8 years comes around U$ 25,000.00 per year. This is the average annual sum of fellowships granted to the students and technicians allocated to the E-Foto project. It is also worth mentioning the academic context on which the project is taking place, the Master's Program in Geomatics at that university. This academic environment has permitted the necessary integration of different kind of scientists and researchers, as well as attracted undergraduate and graduated students. In fact the photogrammetric modules developed by the E-Foto team have being used in the last five years not only in Brazilian Digital photogrammetry Syllabi (at both Masters and Undergraduate levels) but also all over the world. For the sake of exemplification, we highlight the statistics about The E-Foto Homepage views and visits since its creation, in September 16, 2004, which sum 225,394 and 72,242, respectively. Fig. 6.2 presents the number of page visits in September 2010.

Fig. 6.3 shows another relevant information, as far as the geographical distribution of E-Foto Homepage all over the world is of concern.

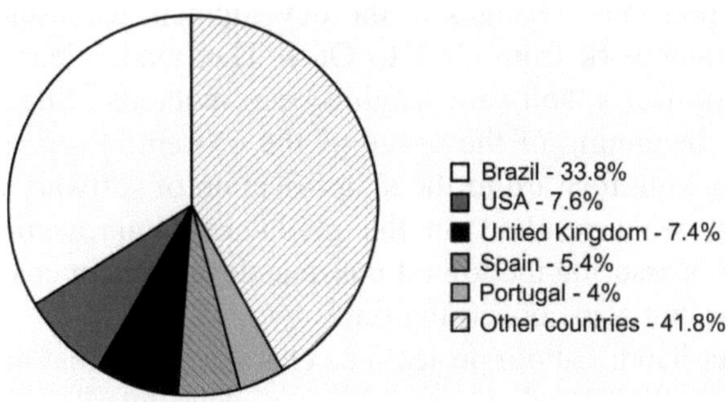

Fig. 6.3. Geographic distribution of the E-Foto Home page views and visits in September 2010

Those who are interested in further details about the development environment are invited to visit the E-Foto homepage, available at http://www.efoto.eng.uerj.br.

6.3.1 Software Development Platform

The project's original proposal means that the E-Foto DPW must be multiplatform, free of proprietary software, adequate to both scientific and graphic software development. Therefore, the software development platform should underlie these requirements. The choice of C++, as the programming language, is due to its usage in several scientific software development projects. Indeed, ANSI C++ is used indistinctly in E-Foto. GUI programming is supported by Qt (Blanchette and Summerfield 2008), while OpenGL and SDL underlie E-Foto's imagery visualization framework.

6.3.1.1 Software Architecture

The E-Foto DPW software's architecture employs Model–View–Controller (MVC) pattern (Gamma et al. 1997). The main reason for the choice of MVC is benefiting from the isolation of problem-domain's logic from control and input/presentation activities. Among the advantages are: independent development, testing and maintenance of each package, subdivision of tasks and the development of specialized competences inside programmers team as well as permitting the concentration of efforts on digital photogrammetry which is the most risky DPW's component.

6.3.1.2 Extreme Programming Practices on E-Foto's Development

E-Foto's productivity, in terms of software development, came down, in 2008, due to personnel changes in the development team and porting E-Foto's GUI framework from Qt 3 to Qt 4. That made clear the need for revision on project's software engineering methods. Such drawbacks motivated the beginning of the usage of the eXtreme Programming (XP) philosophy, looking forward to the simplification of software maintenance and extension, what resulted on the modularity improvement. Another important aim is assuring the lowest possible development cost, not only in financial terms, but also concerning time.

On the other hand, E-Foto project has characteristics that are suitable to XP, such as parallel development of subprojects; interdisciplinary team; short development cycles, of at most two years (usually one); on-the-fly software requirements changes; and permanent rotation of personnel. Adaptation to these adversities is part of the requirements of E-Foto's team. In addition, it is important to create conditions to make new participants adapt quickly, keeping team's productivity unaltered.

On software engineering terms, XP is an agile development methodology (Shore and Warden 2008), which has been widespread with a reasonable amount of success, in the last few years. Based on the initial four principles of communication, simplicity, feedback and courage, and the later added fifth principle of respect, XP promotes high client-developer integration, while reducing the bureaucratic steps impairing this communication.

Some people take XP's practices as really "extreme", such as understanding that there will be no documentation when it is said that working software has more business value than extensive documentation. However, when well applied, XP's practices, such as pair programming, successive refactoring, unity testing by functionality and the establishment of more effective communication ways, can bring significant improvements to the development process. That is the case of E-Foto, where embracing these techniques greatly contributed to DPW's integration and to an outstanding level of algorithm efficiency.

6.4 Implementation of Subprojects

The software developed by E-foto project evolved to a modular architecture in order to permit the concomitant development of distinct parts, hereafter referred by the term subprojects.

6.4.1 Independent Modules

The first cycle of E-Foto DPW implementation produced independent educative modules which were brought about to give support to digital photogrammetry undergraduate and post-graduation courses.

According to (Coelho Filho and Brito 2007), the core of topics related to digital photogrammetry syllabus comprises the following: (1) introduction to photogrammetry; (2) basic principles of photogrammetry; (3) digital image processing applied to photogrammetry; (4) interior orientation of digital imagery; (5) exterior orientation of digital imagery; (6) analytical photo-triangulation; (7) image rectification and normalization; (8) digital surface model extraction; (9) orthorectification, and; (10) photogrammetric stereo-plotting.

A comparison between the syllabi of a reviewing course and three Master Courses in digital photogrammetry with the modules of the Educational DPW reveals that most of the parts of the core subjects in digital photogrammetry are covered. Thus, such educational environment permits students to analyze the didactic contents of the photogrammetric topics, to manipulate and to establish relationships between variables and the performance of the algorithms, as well as to become familiar to the mathematical principles behind the implemented solutions through the source code examination. The study of the orthorectification, for instance, aims at transforming central-perspective images onto orthoimages. According to E-Foto's pedagogical philosophy, there is no sense in introducing both the central perspective and the orthogonal one without explaining the mathematical principles and algorithms underlying the whole process.

6.4.2 System Analysis

To give support to digital photogrammetry teaching, E-foto's independent modules development focused the demonstration of photogrammetric principles following the way done in proprietary DPWs. The short time to which these results were due, in addition to the very small size of the development team at that time, postponed the necessary software analysis.

6.4.2.1 Photogrammetry's Conceptual Model

DPWs are complex applicative software whose development demands careful system analysis. As starting point, the analysis should focus the domain-specific model, leaving aside the view and control abstraction layers. In that context, the analysis exclusively concerns the problem domain, which, in this case, is digital photogrammetry.

Unified Modeling Language (UML) underlies the model's package development of E-Foto DPW, process that can be split up in two steps. First, a conceptual model is done independently of the implementation details, such as concurrency or data storage. Then, it is derived the physical model, figuring out and bringing solutions to more concrete implementation details.

The analysis opened with use cases identification and documentation. A total of ten use cases were identified: (1) open project; (2) images attachment, (3) interior and (4) exterior orientation parameters estimation, (5) bundle block adjustment[3] (6) DTM Generation, (7) stereopair normalization, (8) orthorectification, (9) stereo-plotting, (10) estimate of quality (Coelho Filho and Brito 2007).

After the definition of the use cases, it was necessary to identify the entities involved in each of them, that are, in most cases, domain specific classes taking part in the conceptual model. An additional issue corresponds to the identification of the UML relationships between the domain classes. That approach produced the original E-Foto's DPW object oriented conceptual model. Then, after the identification of the design patterns (Gamma et al. 1997) and aspects (Rashid and Mehmet 2007), two additional conceptual models were derived. The previously mentioned conceptual models were compared considering function point analysis (Fetcke et al. 1998). The results indicated design patterns as the best choice for the development of the physical model of E-Foto DPW.

[3] Bundle block adjustment is a procedure that leads to the exterior orientation parameters of a group of images at once.

6.4.2.2 Photogrammetric Data Model

Project files, in the DPW environment, incorporate input data and the outcomes of the different steps of the photogrammetric workflow. Besides storing information, the project file underlays the data exchange and, therefore, gives support to the integration among the different modules of a DPW. In the case of E-Foto, the development team decided to derive an eXtensible Markup Language (XML) based language, so-called E-Foto Photogrammetric Project (EPP). By and large, XML (Harold and Means 2004) can be understood as a tree data model, since it has a single root element. The two key concepts in XML are elements and attributes. In the case of EPP, elements are used to store data in a hierarchic fashion, while attributes are reserved to metadata, primary keys and external keys. The direct descendants of the EPP root element concern such a macro stages of the photogrammetric workflow: project header, terrain, sensors, flights, points, photogrammetric blocks, stereopairs, interior and exterior orientation parameters estimation, as well as normalization, stereoplotting, DEM generation and orthorectification.

If compared to a normal *ad hoc* ASCII, the choice of XML brings some important advantages. First, all sources of information in the photogrammetric project, including project estimated parameters, images, results and quality measurements are treated indistinctly. Besides, project auditing in E-Foto DPW can be performed by simply opening the EPP in an Internet browser. Moreover, if appropriately conjugated to an eXtensible Stylesheet Language Transformations (XSLT) script, the EPP can be presented in accordance to the level of detail demanded by the user.

6.4.3 Stereo-plotting Cartographic Features to the Brazilian Spatial Data Infrastructure

In the early 1990s topographic map production in Brazil shifted from the traditional analog photogrammetric map compilation to the computer-aided mapping production. This change of paradigm brought many practical consequences and problems. Its main drawback was the lack of standardization of cartographic mapping data structures and patterns. To overcome such a problem, the National Commission of Cartography (CONCAR) launched, in October 2007, a pattern composed of standard specifications for topographic mapping production. This is what is meant by the "Structuring of Digital Geospatial Vector Data" (EDGV, acronym composed by its Portuguese name initials). In 2008, it was published the Decree number 6666 that established the National Spatial Data Infrastructure and ratified the use of standards approved by the CONCAR.

This document brings as attachment the EDGV object oriented conceptual model.

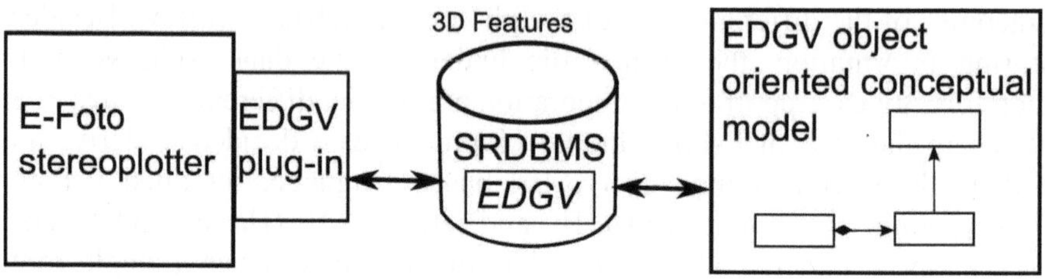

Fig. 6.4. Structuring of Digital Geospatial Vector Data (EDGV) plug-in

As far as the E-Foto DPW is open source software, its research team got involved with an additional technical challenge: the development of a plug-in for direct structuring the 3D geospatial data generated by the photogrammetric map compilation process in the E-Foto environment into the EDGV standards (Borba 2009). That research aimed to provide an implementation of a prototype to be used to photogrammetric compilation of topographic mapping in Brazil. As it is shown in Fig. 6.4, the approach was the derivation of a physical model of the EDGV onto spatial relational database management systems (SRDBMS). One important requirement to this plug-in is flexibility in order to simplify its adaptation to future releases of EDGV. Indeed, the plug-in is able to adapt to any geospatial structure in the SRDBMS.

As a final remark, it is worth mentioning that, by the time of the writing of this chapter, there are some efforts from the Brazilian Government to gather financial support to the E-Foto team. This funding aims the adaptation of the stereo-plotter module of the E-Foto to the photogrammetric map compilation. The stereo-plotter module is sought to be used in topographic map production for public and private cartographic mapping companies in Brazil.

6.4.4 Imagery Visualization Framework

E-Foto uses Qt4 as its GUI's framework, what originally included images visualization. However, the images' visualization system of Qt4 does not give the necessary portability to E-foto, since it uses proprietary software. For that reason, the Linux operating system cannot give total support to it. In consequence, the visualization system has a poorer performance on Linux than on Windows operating system. This was the main motivation for the development of E-Foto's imagery visualization framework. That framework was developed in C++ programming language, using the OpenGL computer graphics API and the SDL multimedia library.

On one hand, OpenGL is being used for the visualization of images, filtering, application of zoom and pan and for marking points on images. On the other hand, SDL is being used for loading the image files and splitting them into pieces so that OpenGL can generate the respective textures. This is necessary since E-Foto DPW employs high-resolution images. Such method allows E-Foto to open any image on any operational system. It must be mentioned that window's creation and event handling are still responsibility of Qt4, since it is the framework used by E-Foto's GUI.

6.4.5 Software Integration

The educational modules, which were originally independently developed, needed to be integrated. That required E-Foto's DPW architecture revision and its adaptation to the photogrammetric data model and photogrammetry's conceptual model.

The success of the integration process demanded careful analysis of the previously used third-party software and study about the best methods to develop the tool. In addition, that intended to make further unavoidable enhancements easier to carry out. The integration process also brought about a new module able to manage the photogrammetric projects, what made E-Foto DPW full oriented to photogrammetric projects.

6.5 Additional Photogrammetric Commons Products

Besides the products about to be mentioned in the following sections, until the time of writing of this document, the project had brought about four master dissertations, seven undergraduate projects and a number of papers published in scientific conferences.

6.5.1 Photogrammetry Book

As mentioned before, the E-Foto project has earned financial support by the Brazilian Research Council (CNPq). That funding had permitted the edition and printing of 1,000 volumes of a book on digital photogrammetry (Coelho Filho and Brito 2007). That book was written as an evolution of class notes in digital photogrammetry. Those class notes were initially written by Brito (1999) and, then, rewritten by Coelho Filho (2002), in the format of an e-book. It is important mentioning that such e-book is available in the E-Foto's home page.

It should be stressed that it is the first photogrammetry book with a full digital approach written in Portuguese language to be edited in Brazil. It was printed by the official publisher of the Rio de Janeiro State University and can be purchased by cost price. Another relevant fact is its availability to download in PDF format at no-cost.

6.5.2 Tutorials

After the development of the E-Foto DPW, the need of creating tutorials that could help users exploring the applications features deeply was noticed. Such documentation is responsible to give support to the philosophy of freedom and self-learning. The tutorials have a step-by-step approach on each of the seven project modules: Image Rectification, Interior Orientation, Exterior Orientation, Bundle Block adjustment, Normalization, Ortho-rectification and StereoPlotter. If employed alongside with to the book, they allow the comprehension and approximation between theoretical and practical aspects of this theme. In 2010, Portuguese and Spanish versions became available.

6.5.3 E-Foto's Website

E-Foto's web page, http://www.efoto.eng.uerj.br, gathers and gives access to all project products, including scientific publications. That is a mean to achieve the project principles of liberty and self-learning. It also paves the way for improving both communication and cooperation among developers and users community. Such collaborative philosophy is underlain on the Internet, whose existence made plainer the coordination of volunteers around the world for open source solutions development (Nichols and Twidale, 2003).

The project's web page was developed using the most recent W3C Consortium standards, making it accessible through different web browsers. To assist the interaction between developers and users worldwide, this web page is based on Dokuwiki, an open source Content Management System (CMS).

Dokuwiki is PHP based, what is the main reason for its choice, and does not use a database for data storage. All data is stored in plain text files. This simplifies the data backup: a simple shell script running on web server backups the data directory to a remote machine. Also, all pages remains readable outside Dokuwiki.

E-Foto uses a modified version of Dokuwiki to allow multiple language support. Nowadays, E-Foto's web pages gives support to four languages: Portuguese, English, Spanish and French. However, part of the pages is exclusively available in Portuguese, which is the developers' mother language.

The wiki approach follows the collaborative philosophy: as a simple and powerful tool, the wiki allows all E-Foto members, even those without HTML skills, to write web pages. A powerful characteristic of Dokuwiki is called "namespace". Namespace organizes the directory structure of raw pages and the site layout. Besides that, E-Foto web site uses namespaces in order to simplify pages translations. There is a namespace for each language supported. When a collaborator wants to translate a page, he/she only creates that page with the same name of the original one, but prefixed with the namespace of the target language and colons. For example: the main page is pt-br:start for the Brazilian Portuguese and en:start for the English main page. When a visitor arrives at main homepage, the wiki tries to guess the browser original language and the correct page is rendered automatically. The other namespaces are es:, for Spanish, and fr:, for French.

E-Foto also uses the flexible access control of Dokuwiki to control which page each collaborator can have permissions to modify.

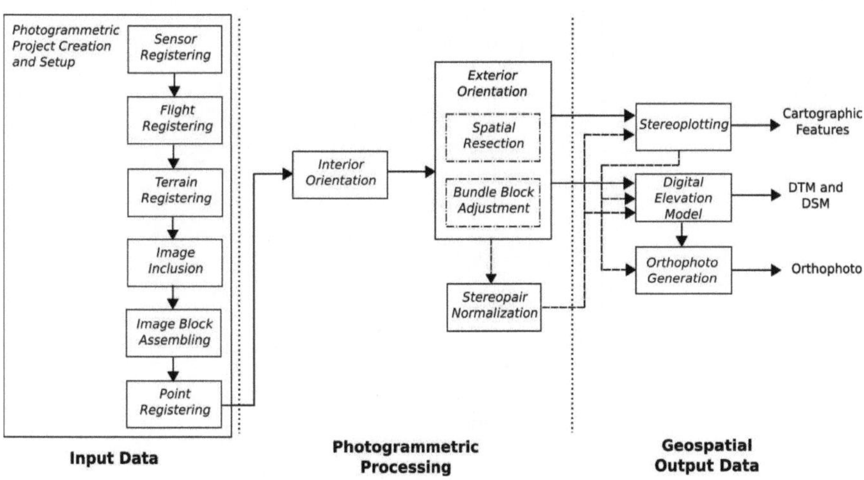

Fig. 6.5. Workflow in E-Foto integrated DPW

6.6 Current State of the Art of the E-Foto DPW

The previously mentioned E-Foto DPW integration aims to turn it oriented to photogrammetric projects. Fig. 6.5 presents an overview of the workflow of E-Foto DPW. Through the decomposition in horizontal direction of this figure, the workflow can be divided into three groups: (1) input data registering and photogrammetric project setup; (2) photogrammetric processing and estimation of images' orientation parameters; and (3) geospatial output data. Rolling back to the past, the independent educative modules of E-Foto DPW comprises seven modules, each one designed to teach one step of photogrammetric process needed to operate the previously mentioned groups (2) and (3).

On the other hand, the first integrated version of E-Foto DPW, whose GUI's screenshot is presented in Fig. 6.6, encompasses group (1) of the photogrammetric workflow, interior orientation and spatial resection, that are part of group (2). Concerning group (3), that release still has a weak integration, what means the adaptation and reuse of the independent modules to make them accept as input the XML photogrammetric data

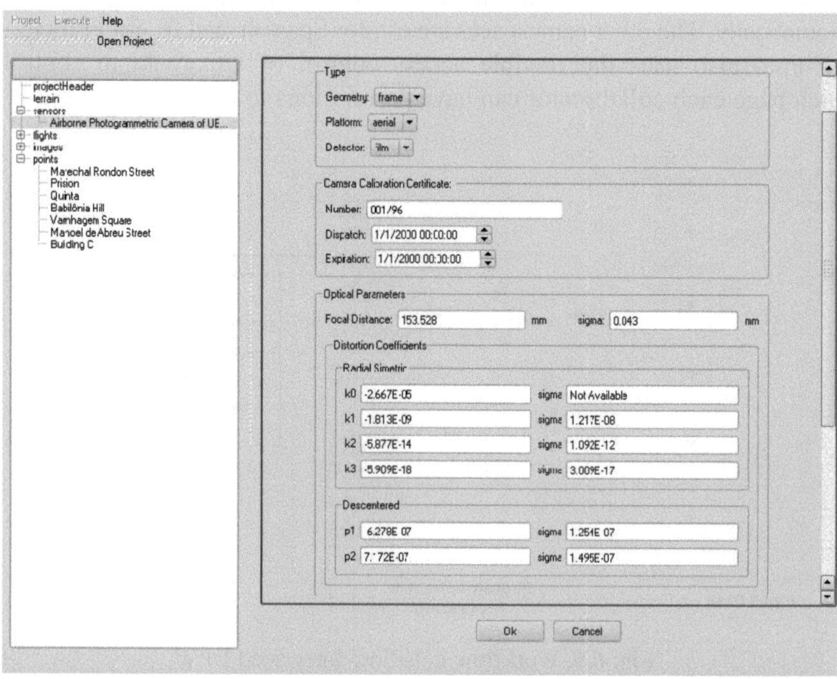

Fig. 6.6. Screenshot of E-Foto DPW while presenting sensor's data

model. The remaining modules will be gradually integrated to E-Foto DPW through successive releases. Regarding sensors, such version was designed to only give support to aerial analogical frame photogrammetric cameras. By the time of the writing of this chapter, the support to additional sensors is being implemented.

Conclusion

By the time of the writing of this chapter the E-Foto project is ongoing in the Master's Program in Geomatics at Rio de Janeiro State University, context in which the E-Foto DPW integration is taking place. Such a work is bringing about the possibility to the user to work with any aerial digitized-frame images, to create a photogrammetric project, and to perform photogrammetric functionalities, such as image orientation, DTM extraction, orthorectification and stereo-plotting, in a single environment. This version aims at overcoming the two major limitations of former versions of E-Foto software: the lack of workflow and the only possibility of working with the imagery of The Rio de Janeiro State University Campus.

We can envision a fruitful future for the E-Foto project and its resulting software. For example, the accesses and page views of the E-Foto project by a number of 252 countries along the time is a very encouraging fact. The E-Foto team has also been received many e-mail messages of congratulations for its initiative.

As a final word, the E-Foto development team is finishing the first integrated version of the E-Foto DPW software. Such DPW is available in the sourceforge SVN environment what permits to really implement the collaborative software development approach. In this approach developers all around the World may contribute to the E-foto project and people (students, teachers and professionals) could benefit of this open source photogrammetric tool.

References

ASPRS (1997) Digital photogrammetry: an addendum to the manual of photogrammetry. The American Society of Photogrammetry and Remote Sensing, Bethesda, Maryland

Blanchette J, Summerfield M (2008) C++ GUI programming with Qt 4, Prentice Hall, Westford

Borba RLR (2009) Restituição Fotogramétrica para a Mapoteca Nacional Digital em Ambiente de Software Livre, Mastering Dissertation. The Rio de Janeiro State University. The Masters Program in Geomatics

Brito JLNS (1999) Class notes on digital photogrammetry. Brazilian Military Institute of Engineering, Rio de Janeiro, RJ

Coelho Filho LCT (2002) Projeto E-FOTO: Uma Estação Fotogramétrica Digital Educacional. Undergraduate project, Brazilian Military Institute of Engineering

Coelho Filho LCT, Brito JLNS (2007) Fotogrametria Digital. Eduerj- The Rio de Janeiro State University Press, Rio de Janeiro, RJ

Fetcke T, Abran A, Nguyen TH (1998) Mapping the OO-Jacobson approach into function point analysis. Technology of object-oriented languages and systems. IEEE Comput Soc, Los Alamitos

Gamma E, Helm R, Johnson R, Vlissides JM (1997) Design patterns: elements of reusable object-oriented software. Addison-Wesley Professional, Reading

Harold ER, Means WS (2004) XML in a Nutshell. O' Reilly Media, Sebastopol

Heipke C (2001) Digital photogrammetric workstations – a review of the state-of-the-art for topographic applications. GIM Intl 15:4, 35–37

Kraus K (2000) Photogrammetry. Dümmler, Köln

Mikhail EM, Bethel JS, McGlone JC (2001) Introduction to modern photogrammetry. Wiley, New York

Nichols D, Twidale B (2003) The usability of open source software. First Monday 8:1

Rashid A, Mehmet A (2007) Transactions on aspect-oriented software development IV. Springer, Berlin

Schenk T (1999) Digital photogrammetry. TerraScience

Shore J, Warden S (2008) The art of agile development. O'Reilly, Sebastopol

Chapter 7 An Overview on Interferometric SAR Software and a Comparison Between DORIS and SARSCAPE Packages

Elisabeth Simonetto, Jean-Michel Follin

Ecole Supérieure des Géomètres et Topographes, 1 Bd Pythagore, Campus Universitaire, 72000 Le Mans, France,
{elisabeth.simonetto, jean-michel.follin} @esgt.cnam.fr

Abstract. Synthetic Aperture Radar (SAR) Interferometry, or INSAR, relies on the processing of SAR data. INSAR relies on the processing of SAR data leading to Digital Elevation Models (DEM). DINSAR (Differential INSAR) or A(dvanced)-DINSAR provide ground deformation measurements. These techniques have proven their capabilities for several years but they are generally only used by experts in radar image processing. However, several free-of-charge (freeware or open-source) and proprietary software packages exist. In the meantime, new SAR satellite sensors are launched, which improves the availability of SAR data with pertinent acquisition dates and viewing parameters. In this context, a brief introduction to these software packages and SAR data is given from the end-user point-of-view with an emphasis on their gratuitousness and respect of open-source concepts. From that work, the open-source processing package DORIS[1] was chosen for the research activities at ESGT. In order to validate our choice, DORIS is studied in the framework of a DINSAR application. The software organization and the obtained results are cross-compared to those obtained with a second proprietary software, SARSCAPE. Experiments are carried out with satellite radar data acquired before and after the earthquake that occurred in Bam, Iran in 2003. The obtained results show differences that can be explained by different methods used to refine the orbital information in DORIS and in

[1] Delft Object oriented Radar Interferometric Software

E. Bocher and M. Neteler (eds.), *Geospatial Free and Open Source Software in the 21st Century*, Lecture Notes in Geoinformation and Cartography,
DOI 10.1007/978-3-642-10595-1_7,

SARSCAPE. These differences do not prevent us from validating our choice and our use of DORIS as outcome of this comparison.

7.1 Context and Objectives

7.1.1 Introduction of Two-Pass DINSAR

In this paper, we focus on two-pass DINSAR (Differential Interferometry of Synthetic Aperture Radar data) using satellite SAR images. Two-pass DINSAR aims at providing ground deformation measurements from two SAR data, acquired by the same sensor during the flying over the same area at two different times. For almost twenty years (Gabriel et al. 1989), a variety of studies have shown the efficiency of DINSAR to detect and survey vertical ground deformation with annual displacement rates from millimetres to a few centimetres (Colesanti and Wasowski 2006; Crosetto et al. 2005). In favourable cases, the classical technique provides measurements with a good precision (mm/yr) over large areas (100 × 100 km) with typical ground resolution of 20 m. Measurements are performed in the Line-Of-Sight (LOS) direction of the sensor and usually show high sensitivity to the vertical component of the deformation field. Several applications of DINSAR are known: urban monitoring, post-mining monitoring, civil engineering, gas or petrol field surveillance, aquifer system analysis, earth science (seismic activity, volcano, glacier), etc. However, the application of DINSAR depends on several factors, such as soil nature and its geometry, deformation characteristics (extend, amplitude, direction), sensor parameters (wavelength, spatial resolution, viewing angle), and the availability of suitable SAR data.

A SAR sensor is an active remote sensing system. It emits a beam of microwaves and captures the backscattered echoes from the illuminated surfaces (Oliver and Quegan 1998). A bidimensional signal is registered. The first dimension is the azimuthal axis and the second one is the target-antenna or range axis. SAR processing (or focusing) of this signal leads to the Single Look Complex (SLC) image. Each pixel has a complex value: amplitude and phase. This phase is not known in an absolute way but is measured modulo 2π. The only phase is unexploitable because it varies too fast between nearby pixels. On the other hand, the difference of phase between the echoes acquired from two close orbits varies more slowly. This difference in phase, computed for each pixel, is the interferogram: it provides information on topography and ground motion. INSAR computes a Digital Elevation Model (DEM) from this interferogram, assuming no deformation. Two-pass DINSAR retrieves the deformation signal from the interferogram and a coarse DEM. However, the interferogram is also influenced by other factors such as atmospheric signal and inaccuracies of the orbital information.

The processing of the two SAR images consists in several steps. Firstly, both datasets are first coregistered (Fig. 7.1). The first SAR dataset, the master dataset, is used as geometric reference. The second one is the slave dataset. The interferogram is computed and filtered using spectral shift filterings (adapted filtering based on signal processing theory).

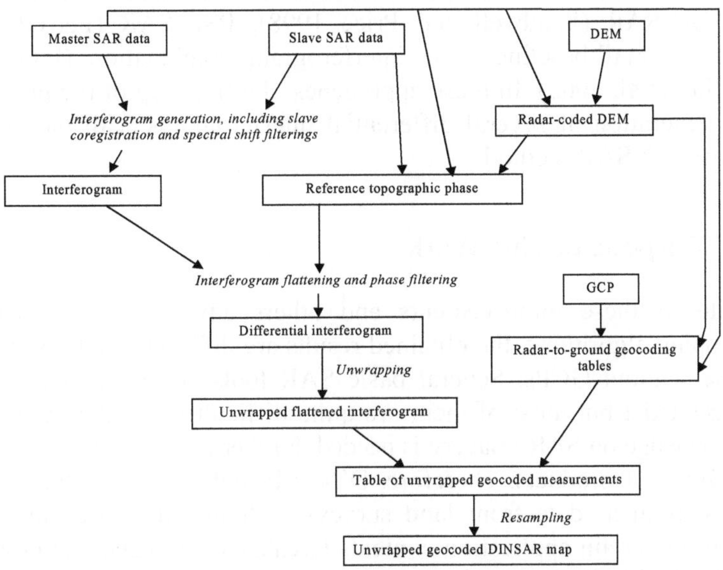

Fig. 7.1. Two-pass DINSAR flowchart

A coherence map is also produced (a map indicating for each pixel the correlation degree between the two SAR signals). Then, to isolate the deformation signal, the topographic component is deleted using a coarse DEM. For that purpose, the DEM is radar-coded (resampled according to the master) and used to generate a reference topographic phase. The differential interferogram is obtained by subtracting the reference topographic phase from the interferogram (interferogram flattening). It appears as fringes, each fringe corresponds to the phase variation of one cycle and can be interpreted as deformation contour lines, separated by a distance equal to half of the radar wavelength. The differential phase is an ambiguous measure of the ground displacement: it is wrapped. Thus, the unwrapping processing aims at determining the unambiguous phase (or unwrapped phase), that is proportional to the vertical ground deformation. DEM and Ground Control Points (GCP) are then used for the geocoding of the map.

As mentioned by Colesanti and Wasowski (2006), over the last 10 years advanced signal processing techniques have been proposed to overcome some limitations of the classical two-pass DINSAR approach. They are denoted as A-DINSAR by Crosetto et al. (2005). The methods aim at isolating the ground deformation signal from the other contributions (topography, atmosphere, orbit) and rely on the exploitation of many SAR images covering a long time period. The following methods can be found: Stack DINSAR (Sandwell and Price 1998), PSINSAR (Ferretti et al. 2000), or small-baseline (SB) interferogram combination (Usai 2003, Berardino et al. 2002). In these approaches, the first step of the processing is the generation of several differential interferograms with the classical two-pass DINSAR method.

7.1.2 Purpose of This Work

In spite of these improvements and others advancements concerning unwrapping algorithms, the obtained results are difficult to interpret and to validate without GCPs. Several basic SAR tools and DINSAR software packages exist but most of them are quite difficult to handle by a novice and knowledge on SAR imagery is needed. Furthermore only few companies offer DINSAR products. Therefore, DINSAR methods or products are not used as often as data from land surveys, although they are particularly efficient in certain application contexts (weakness of vegetation coverage, etc.). With this background, we aim at giving an overview on existing software and DINSAR products. We think that free-of-charge and open-source programs could enable a real expansion of the use of these techniques by anyone which is interested in DINSAR. Free-of-charge programs can be easily tried in reason of their gratuitousness. They include first open-source software which allows end-user to keep the control on the software functioning (advanced parameters setting, algorithm implementation, …) and enables addings, and secondly freeware ones which do not. For this reason, in this work, we will consider two categories of software: "free-of-charge and sometimes open-source" ones, and "proprietary and commercial" ones. The processing suite DORIS was chosen for different reasons as it will be discussed in the following of the paper. In particular, it is both free-of-charge and open-source under certain restrictions (for non-commercial use). In order to validate our choice and our use of DORIS, we compared DORIS results to results obtained with the proprietary and commercial software SARSCAPE. The paper continues as follows: in Sect. 7.2 a review of available SAR data and software is given. In Sect. 7.3 we introduce the DORIS package. In Sect. 7.4 we present the results obtained with DORIS and SARSCAPE and underline their differences.

7.2 An Overview on Radar Sensors and Software

7.2.1 SAR Sensors

Synthetic Aperture Radar sensors can be on board of airplanes, shuttles or satellites. They are manufactured by different institutes and data are generally distributed by the concerned space agency. Some data are free-of-charge for research purposes. To our knowledge, satellite SAR data have open format specification. The SAR systems distinguish from each other in wave characteristics, antenna agility[2] and system performance in terms of repeat cycle[3] and spatial resolution. For the sake of brevity, the discussion is here limited to spatial systems. The first satellite SAR sensor was aboard the American SEASAT satellite launched in 1978. INSAR applications have shown a greater rate of use with the launch of the ERS1 satellite of the European Space Agency (ESA) in 1991. ERS1 was followed by ERS2 in 1995 and ENVISAT in 2002. In 2000, the Shuttle Radar Topography Mission (SRTM) acquired SAR data all around the world within 11 days. Based on that data a freely downloadable global DEM has been produced with INSAR, with a 3 arc-second pixel resolution and an absolute vertical accuracy of about 10 m. One can also utilize data from the Canadian RADARSAT1&2 (RDS1&2), launched in 1995 and 2007, the Japanese JERS (launched in 1992) and ALOS (launched in 2006), or from the Indian RISAT1 (launched in 2007). Besides those projects a variety of other projects have been announced that will deliver data of high spatial resolution (up to 1 m), with high repeat cycle, with different polarimetric wave modes, and more suitable geometric configurations for interferometry. These systems are made up of several satellites. For example the Italian Cosmo-Skymed (CSK) constellation of four satellites is built since 2007. The German TerraSAR-X (TSX) was launched in 2007 and will be followed by Tandem-X. The Argentine SAOCOM1 will be made of two satellites. This listing is not complete (Rosen and Buccolo 2007) but shows that a consistent

---

[2] The antenna can move to modify the look direction.

[3] The repeat cycle is the temporal base between two successive passes of the satellite over the same area allowing an interferometric measurement.

SAR data archive is available and that SAR data acquisition will be also performed in the future.

7.2.2 Interferometric Software

In this paper, we distinguish two kinds of software: packages and radar tools. A package is an integrated suite of tools. A two-pass DINSAR software package consists of an ordered sequence of data treatments leading to the geocoded ground deformation map from two SAR images and a DEM. It is also named an interferometric processing suite. Radar tools are software containing various functionalities, each of them dedicated to a particular task like data reading or noise filtering.

We deem the following criteria as important in the software selection process:

- Costs: free-of-charge or commercial
- Existence and cost of maintenance, hotline or mailing list
- Update frequency
- Platform portability: Windows, Linux, Unix, MacOSX
- Software capabilities: processing (SAR focusing, unwrapping, geocoding, automatic improvement of orbit parameters, resampling tools, etc.), supported data formats and output options (product format, supported geocoding systems)
- Ergonomics and documentation quality: according to the user knowledge in computer science and SAR processing techniques
- Access to source code

Among these criteria, we think that the gratuitousness and the access to source code are important because it encourages the use of DINSAR, enables users an advanced study of the DINSAR processing functions and allows modifications and extensions, in particular for experimenting data processing methods. For these reasons, we propose a listing where we choose to separate proprietary-and-commercial software from free-of-charge ones (Tables 7.1 and 7.2). The free-of-charge products that are presented in the Table 7.1 are not necessarily and their use can be restricted (generally to scientific purposes). Here, open-source means that the source code is accessible. In the tables, we also present the basic capabilities for each software and their license status. Our listing is not exhaustive since it is sometimes difficult to find reliable information. But the tables aim at giving to a potential SAR software user a starting point for a further detailed software analysis. Let us note that previous listings have been proposed in Crosetto et al. (2005) and Gens (1999).

Nowadays, we can find five main proprietary-and-commercial packages: SARSCAPE (Sarmap), GAMMA (Gamma Remote Sensing), DIAPASON (CNES), IMAGINE radar Mapping Suite (Geosystems) and EarthView (MDA). They allow the complete DINSAR processing, sometimes the data focusing and A-DINSAR techniques (in SARSCAPE, GAMMA, EarthView). Their main advantages are ergonomics, documentation, hotline, and maintenance with regular and frequent updates, platform portability. These updates enable the assimilation of data acquired by the last launched SAR systems. To our knowledge, the source code is only available with the Gamma software.

Four main free-of-charge packages exist (Table 7.1): ROI_PAC, DORIS, RAT and STAMPS. ROI_PAC and DORIS allow the two-pass DINSAR processing. Concerning their interfaces, ROI_PAC, DORIS and STAMPS are command line based packages. RAT is more appropriate to INSAR but has a graphical user interface (GUI) that simplifies its use. STAMPS is dedicated to A-DINSAR (PS technique) but uses ROI_PAC for focusing and ROI_PAC or DORIS to compute the differential interferograms. Concerning the license status, ROI_PAC is protected by a license that limits its scope to private use and bans its redistribution: commercial use should be made under a license contract. Despite the possibility to redistribute or modify DORIS under the terms of a GNU GPL,[4] conditions of use for scientific purpose are added in DORIS. RAT is under a MPL[5] and STAMPS is under an unknown open-source licence.

Several tools exist for reading, displaying, resampling, noise filtering, geocoding, etc, SAR data like: ENVIVIEW, NEST, OTB and ASF SAR tools. Their main capabilities, related to a DINSAR processing, are presented in Table 7.2. Since recently, NEST, under a GNU GPL, includes a plugin for the interferometric processing with DORIS and is made with a GUI. OTB is under a particular French license: CeCILL[6] and the associated GUI, named Monteverdi, allows an easier use of some of the OTB functions. POLSARPRO is focused on SAR data acquired in the quad polarimetric mode and allows their Pol-INSAR processing. It is also made up with a GUI. As RAT, IDIOT is well adapted for a non-expert using ENVISAT data in a DINSAR context but it does not yet lead to a geocoded product. STUN is made of Matlab tools devoted to PSINSAR. SNAPHU and GETORB are used in DORIS. Copyrighted SNAPHU aims at unwrapping and open source GETORB scope is to retrieve precise satellite orbits.

---

[4] GNU General Public License.

[5] Mozilla Public License.

[6] Cea Cnrs Inria Logiciel Libre.

We can notice that Geospatial Data Abstraction Library (GDAL) is an open-source translator library for raster geospatial data formats offering readers for certain SAR data like for instance CEOS[7] SAR image files.

Thus there are several alternative. We choose DORIS because of its advantages: open-source and free-of-charge aspects (for scientific use), capabilities and interoperability (complete INSAR processing, used by STAMPS, recently chosen by ESA in NEST), conviviality (hotline possible via a mailing list, good documentation, easiness of use). However, some drawbacks can be highlighted: no GUI (this may be resolved with NEST), irregular updates, all satellite format are not supported, no support for aerial sensors. In the next section we will concentrate on DORIS, and will later compare the results obtained with DORIS with result obtained from SARSCAPE.

Table 7.1. Free-of-charge INSAR software packages (in April 2009)

Name	Company	Capabilities under interest	Supported SAR satellite sensors	Release	Platform	License
ROI_PAC (Rosen et al. 2004)	JPL / Caltech	SAR focusing, INSAR, DINSAR	JERS1, RDS1&2, ERS1&2, ENVISAT, ALOS	3.0 (2007)	Unix / Linux	Open Channel Software / Caltech
DORIS (Kampes and Usai 1999)	TU DELFT	INSAR, DINSAR, no unwrapping	JERS1, RDS1&2, ERS1&2, ENVISAT, ALOS, TSX	3.20 (2008) 4.02 (06/2009)	Unix / Linux / MacOSX	Redistri-bution and/or modification under GPL v2 or later version
RAT (Reigber and Hellwich 2004)	Berlin Univ. of Techno-logy	INSAR, POLINSAR, no geocoding	JERS1, RDS1&2, ERS1&2, ENVISAT, ALOS, TSX	0.20 (2008)	IDL VM (Linux / Unix / Windows / MacOSX)	MPL 1.1
STAMPS (Kampes 2005)	Stanford Univ.	PSINSAR	JERS1, RDS1&2, ERS1&2, ENVISAT, ALOS	2.2 (2007)	Unix / Linux	Unknown, open-source

[7] The CEOS format is a standard format of SAR data that was defined by the CEOS (Committee on Earth Observation Satellites) international working group subgroup on SAR format standardisation in 1989.

Table 7.2. Other free-of-charge radar tools (between April 2009 and January 2010)

Name	Company	Capabilities under interest	Supported SAR satellite sensors	Release	Platform	License
ENVIVIEW[8]	ESA	Basic tools for SAR data handling	ERS1&2, ENVISAT	–	Unix / Linux / Windows	Unknown (Freeware)
NEST[7] (following BEST)	ESA	Basic tools for SAR data handling	JERS1, RDS1, ERS1&2, ENVISAT, ALOS, TSX	2C-1.2 (06/2009)	Unix / Linux / Windows / MacOSX	GPL v3
OTB[9]	CNES	Basic tools for SAR data handling	ERS1&2, ENVISAT, RDS1&2, ALOS, TSX, CSK	3.2.1 (2010)	Linux / Windows / Unix / MacOSX	CeCILL
ASF SAR tools[10]	ASF	In particular SAR Processor	Not found	1.1.10 (2009)	Windows/ Linux	Unknown
POLSAR-PRO (Pottier et al. 2005)	ESA / Rennes-1 Univ.	Polarimetric tools among which POLINSAR	SIRC, RDS2, ENVISAT, ALOS, TSX	4.0 (2009)	Windows / Unix / Linux	GPL v2
IDIOT (Reigber et al. 2007)	Berlin Univ. of Techno-logy	DINSAR, no unwrapping, no geo-coding	ENVISAT	1.3 (2008)	IDL VM (Linux / Windows / Unix / MacOSX)	Unknown (Freeware)
STUN (Kampes 2006)	DLR	Tools for PSINSAR	–	2006	CDROM	Unknown, open-source
SNAPHU (Chen and Zebker 2002)	Stanford Univ.	Unwrapping software	–	1.4.2 (2003)	Unix / Linux	Unknown, copyright
GETORB (Scharroo and Visser 1998)	TU DELFT	Precise orbit retrieval	ERS1&2, ENVISAT	2.3.2 (2008)	Unix / Linux	Unknown, open-source

8 http://earth.esa.int/resources/softwaretools/, accessed 05 January 2010.

9 http://www.orfeo-toolbox.org/otb; accessed 05 January 2010.

10 http://www.asf.alaska.edu/sardatacenter/, accessed 05 January 2010.

7.3 Two-Pass DINSAR Processing with DORIS

Interferometric processing consists of several steps (Fig. 7.1). We expect to find the same tools, to accomplish a particular step, in free-of-charge as well as in proprietary-and-commercial software. We will introduce and analyse the software DORIS and compare it with the proprietary software SARSCAPE, focusing only on functionality for two-pass DINSAR processing.

7.3.1 DORIS Processing

DORIS can be used to generate a DEM or a deformation map (Kampes and Usai 1999; Kampes 2005). In the following experiment, we used DORIS 3.20 within CYGWIN, a Linux-like environment for Windows. The setup is simple as long as several open-source software libraries (lapack, fftw), packages (Proj.4, GMT), tools (SNAPHU, GETORB), and free compilers are first installed. The software contains 53 modules organized in three main blocks:

1. SAR data reading, precise orbit retrieval (using GETORB), possible cropping and oversampling, parameter computation for the fine sub-pixel coregistration of the slave image to the master
2. Filtering, slave data coregistration, interferogram generation, radar-coded DEM, reference topographic phase generation from DEM, differential interferogram generation (denoted flattening), possible phase noise filtering and multilooking (adapted radar resampling)
3. Unwrapping (using SNAPHU), geocoding grid computation (optionally with GCP), map geocoding and resampling (using GMT) in WGS84 or UTM

7.3.2 Comparison with the Proprietary-and-Commercial Software – SARSCAPE

The software SARSCAPE 4.0 has be used as a plugin for the ENVI 4.4 Remote Sensing software for Windows. SARSCAPE offers several level of SAR processing through a user-friendly GUI. Besides regular updates, a proper documentation and hotline facilities, the organization of the software with respect to DINSAR processing is similar to the DORIS but differences exist:

- The data assimilation does not use GETORB for ERS data and requires other precise orbital files, distributed by ESA, for ENVISAT images.

- The SRTM DEM, often used as coarse DEM in DINSAR processing, is automatically recovered from the Internet.[11] Its Geoid altitudes are automatically converted into ellipsoidal WGS84 heights. For the radar-coded DEM resampling it is possible to introduce GCP's that allow to correct the orbital parameters of the master dataset.
- The spectral shift filtering (a particular signal processing to improve the signal-to-noise ratio of the interferogram) can use the DEM, which may improve the processing.
- A baseline refinement stage is performed to correct the orbital parameters. This step is based on manually selected GCPs, the unwrapped differential interferogram and the radar-coded DEM. Let us note that a processing for the refined correction of ephemerides has been introduced in DORIS 4.02.
- Unwrapping and geocoding are parts of the software.
- Numerous coordinate reference systems and projections are possible.

The main difference between both software packages relies on the different strategies for the orbital parameter refinement. This could generate visible dissimilarities in the results. Of course, even if a processing step has the same goals, the used methods may be different. Additionally different parameter settings may change the results. Experiments are then needed to show the potential differences in results and to know whether those differences are relevant for further analysis.

7.4 Experiments

The previous part introduced DORIS. Experiments have been carried out to compare DORIS results with those obtained with SARSCAPE. The chosen test site is Bam, Iran, for which SAR data have been acquired before and after the 26 December 2003 earthquake. The DINSAR products for this area show ground vertical deformation and many interferometric products have been published in the literature. Hence, the coherence of the obtained products can be checked as discussed in Simonetto (2008). This area can lead to little noisy interferometric measurements because of the characteristics of its soil and climate.

[11]The default ftp site is: ftp://e0srp01u.ecs.nasa.gov/srtm/version2/SRTM3/

7.4.1 Presentation of the Data Set

Two images from the Bam ESA ENVISAT dataset package are processed (Fig. 7.2a). These data can be obtained at no charge for research purposes. The characteristics are the following ones: track 120, frame 3024, IS2 descending mode, orbit numbers 9192 (03 Dec. 2003, master) and 10194 (11 Feb. 2004, slave).

The free-of-cost 3 arc-second SRTM DEM has been obtained as well. It is provided in WGS84 with the geoid as height datum. Several tiles are required to cover the whole area visualized in the master data: N28E058, N28E059, N29E058, N29E059. The relevant tiles are gathered into a mosaic. Then, heights are converted to ellipsoidal WGS84 altitudes using the free-of-charge NGA/NASA 15 min worldwide geoid height file, EGM96. The DEM altitudes vary from 204 to 3,740 m (Fig. 7.2b) and theoretically, its vertical precision is sufficient for our experiment.

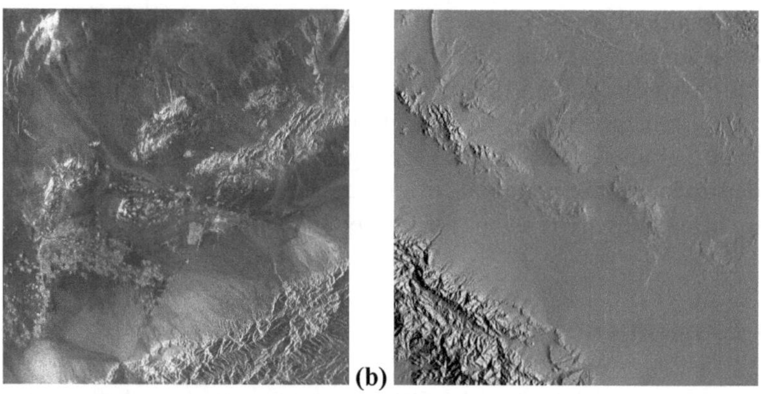

(a) (b)

Fig. 7.2. (**a**) Master amplitude image in radar geometry, Bam (located around 29°10'N in latitudes and 58°20'E in longitudes) is around the center of the scene. (**b**) Shaded SRTM DEM

7.4.2 Main DINSAR Products from DORIS

The radar-coded results given in Fig. 7.3, are satisfactory: the known pattern of the coseismic ground deformation due to the seismicity is recognized (Fig. 7.3a). The unwrapped differential phase is obtained without difficulties (Fig. 7.3b). It shows a maximal uprising of around +30 cm at the bottom left side of the deformation pattern and a maximal subsidence of about –15 cm at the upper left side. For a more detailed analysis on this ground displacement, we refer to Stramondo et al. (2005).

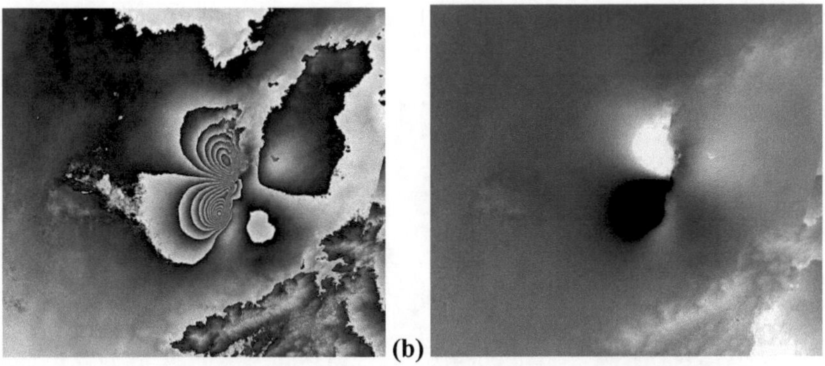

Fig. 7.3. Result of the DORIS processing using a Bam ENVISAT dataset and the SRTM DEM. (a) Differential interferogram. A variation of phase from a bright tone to a dark one corresponds to one cycle, that is: a LOS downward displacement of half of the radar wavelength (here equal to 5.66 cm). (b) SNAPHU unwrapped differential interferogram. The bright area corresponds to an increasing phase, that is, according to the interferogram computation method, a ground subsidence. The black area refers to a decreasing phase, that is a ground uprising

7.4.3 Comparison with the SARSCAPE Results

To enable a comparison, we tried to use the same parameter values in ORIS and in SARSCAPE. The differential interferogram (Fig. 7.4a) shows around two horizontal fringes along the scene in the azimuthal direction. It corresponds to a unwanted residual global slope. We met difficulties to correctly choose GCPs and a better result could probably be obtained. In order to compare the differential interferograms, we choose to first compensate this residual slope with a signal processing and unwrap it using SNAPHU (Fig. 7.4b).

SARSCAPE and DORIS unwrapped products are then subtracted from each other. The difference image shows a global linear slope from the Upper Left corner towards the Lower Right corner. It corresponds to a residual slope of around 5.6 cm along 140 km. This difference is not acceptable as the theoretical measurement accuracy should be here around 2 mm (Zebker and Villasenor 1992). Without GCP, it is however not possible to determine if one result is more reliable than the other one. This trend can be more accurately modeled by a quadratic surface, and is due to residual orbital errors, probably present in both results. When deleting this quadratic trend on both results, we can measure a phase variation in the difference image of about 1.9 cm along 140 km. In this case, for a ground deformation of small spatial extend, e.g. 1 km, the difference between both results become negligible, which allows validating the ground deformation measurements.

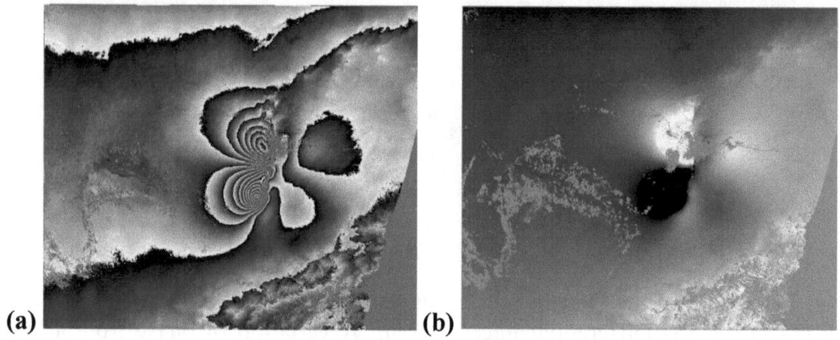

Fig. 7.4. (**a**) SARSCAPE differential interferogram. (**b**) SNAPHU unwrapped differential interferogram. Refer to Fig. 7.3 for explanation of the color tables

7.5 Conclusions and Perspectives

The use of a DINSAR software is neither easy nor impossible for someone without skills in SAR imagery. A software with available source code enables the user to completely understand how the processing is done. In this study, the software package DORIS was examined, which contains all modules to perform a two-pass DINSAR processing, is free-of-cost and open-source. During the handling of DORIS it has been noticed that DORIS is not able to read SAR data acquired with all the newest satellite-based sensors. However, the results obtained for the Bam dataset in Iran are in coherence with the ones computed with the proprietary-and-commercial software SARSCAPE. Differences mainly come from uncorrected orbital inaccuracies. GCPs are required if millimeter precision is needed. The observed differences between the generated results, even weak ones, can confuse the user and may affect his/her confidence, since interferometric measurements should be similar and reproducible if the same data set is used. Only few works (for instance Raucoules et al. 2007) have compared DINSAR results derived with different software. Our results demonstrate that more comparisons, open-source software and free datasets with GCP, even simulated ones, are needed to allow the expansion of INSAR techniques.

Acknowledgments

The authors are thankful to ESA for providing the Bam ENVISAT ASAR dataset package. They thank the developers of the DORIS package for its free-of-charge availability, and the interested parties of the ENVI and SARMAP hotlines for their help on SARSCAPE use. Then, they thank the reviewers who allowed a great improvement of this paper.

References

Berardino P, Fornaro G, Lanari R et al (2002) A new algorithm for surface deformation monitoring based on small baseline differential SAR interferograms. IEEE Trans Geosci Remote Sens 40(11):2375–2383

Chen CW, Zebker HA (2002) Phase unwrapping for large SAR interferograms: statistical segmentation and generalized network models. IEEE Trans Geosci Remote Sens 40:1709–1719

Colesanti C, Wasowski J (2006) Investigating landslides with space-borne synthetic aperture radar (SAR) interferometry. Eng Geology 88:173–199

Crosetto M, Crippa B, Biescas E et al (2005) State-of-the-art of land deformation monitoring using differential SAR interferometry. International society for photogrammetry and remote sensing Hannover workshop. www.ipi.uni-hannover.de/fileadmin/institut/pdf/126-crosetto.pdf. Accessed 03 April 2009

Ferretti A, Ferrucci F, Prati C et al (2000) Nonlinear subsidence rate estimation using permanent scatterers in differential SAR interferometry. IEEE Trans Geosci Remote Sens 38(5):2202–2212

Gabriel K, Goldstein RM, Zebker HA (1989) Mapping small elevation changes over large areas: differential interferometry. J Geophys Res 94:9183–9191

Gens R (1999) SAR interferometry: software, data format and data quality. Photogrammetric Eng Remote Sens 65:1375–1378

Kampes BM (2005) DORIS user's manual and technical documentation. http://enterprise.lr.tudelft.nl/doris/software/doris_v3.20.pdf. Accessed 03 April 2009

Kampes BM (2006) Radar interferometry – persistent scatterer technique. Springer, Dordrecht

Kampes BM, Usai S (1999) Doris: the delft object-oriented radar interferometric software. In: 2nd International symposium on operationalization of remote sensing (cdrom), Enschedes

Oliver C, Quegan S (1998) Understanding synthetic aperture radar images. Artech House, London

Pottier E, Cloude S, Hajnsek I et al (2005) PolSARpro v2.0: the polarimetric SAR data processing and educational toolbox. In: Institute of electrical and electronics

engineering, international geoscience and remote sensing symposium 2005 proceedings 5:3173–3176
Raucoules D, Bourgine B, de Michele M, Le Cozanet G et al (2007) Persistent scatterers interferometry independent validation and intercomparison of results. Final report BRGM/RP-55649-FR http://earth.esa.int/psic4/Documents.html. Accessed 04 April 2009
Reigber A, Hellwich O (2004) RAT (radar tools): a free SAR image analysis software package. In: European conference on synthetic aperture 2004 proceedings, Ulm, pp 997–1000
Reigber A, Erten E, Guillaso S et al (2007) IDIOT: a free and easy-to-use software tool for DINSAR analysis. In: ENVISAT symposium proceedings http://earth.esa.int/workshops/envisatsymposium/proceedings/posters/2P7/463306re.pdf. Accessed 03 April 2009
Rosen PA, Buccolo GM (2007) Perspectives on worldwide spaceborne radar programs. In: Institute of electrical and electronics engineering radar conference, pp 740–745
Rosen PA, Hensley S, Peltzer G et al (2004) Updated repeat orbit interferometry package released. EOS, Trans Am Geophys Union 85(5), doi:10.1029/2004EO050004
Sandwell T, Price EJ (1998) Phase gradient approach to stacking interferograms. J of Geophys Res 103-C4:8113–8127
Scharroo R, Visser P (1998) Precise orbit determination and gravity field improvement for the ERS satellites. J Geophys Res 103:8113–8127
Simonetto E (2008) DINSAR experiments using a free processing chain. In: Proceedings of the SPIE 7109, Cardiff, Wales, doi:10.1117/12.802508
Stramondo S, Moro M, Tolomei C et al (2005) InSAR surface displacement field and fault modelling for the 2003 Bam earthquake (southeastern Iran). J Geodyn 40:347–353
Usai S (2003) A least squares database approach for SAR interferometric data. IEEE Trans Geosci Remote Sens 41(4):753–760
Zebker HA, Villasenor J (1992) Decorrelation in interferometric radar echoes. IEEE Trans Geosci Remote Sens 30(5):950–959

Chapter 8 An Object-Oriented Approach to Automatic Classification of Panchromatic Aerial Photographs with GRASS GIS and R

Eduardo Corbelle Rico, Rafael Crecente Maseda

Laboratorio do Territorio (LaboraTe), Escola Politécnica Superior, University of Santiago de Compostela Laboratorio del Territorio 27002, Lugo, Spain
{eduardo.corbelle, rcrecente} @usc.es

Abstract. Object-oriented classification of high spatial resolution imagery usually offers higher accuracy than traditional per-pixel approaches, as real-world features (objects) appear in those as groups of pixels with a relatively high internal variability – though generally lower than that of the whole image – due to the heterogeneous spectral response of their surface. Although a particular kind of object-oriented classification algorithm is currently implemented in GRASS GIS (the Sequential Maximum a Posteriori estimation or SMAP), it presents several drawbacks when compared to some proprietary alternatives. One of the most important shortcomings of SMAP is the fact that both the segmentation and classification phases have to be carried out on the same group of images: If textural information is used as an additional source of information, this usually increases the confusion between coarse-texture classes and border areas between different covers. Other interesting problem is the absence of any control over the size and shape of the objects created. While these problems might seem enough reason to turn attention to other pieces of software currently available in the market, other alternatives exist: One of such detour approaches (the one presented by this work) might be the use of existing vector layers as segments (objects) which may be classified using statistical techniques such as decision trees. This work presents an example of classification of historical aerial photographs into five broadly defined classes: (1) arable land, (2) pastures, (3) pastures with trees, (4) shrubland, (5) trees. Two different approaches are compared: (a) the standard use of the

E. Bocher and M. Neteler (eds.), *Geospatial Free and Open Source Software in the 21st Century*, Lecture Notes in Geoinformation and Cartography,
DOI 10.1007/978-3-642-10595-1_8, © Springer-Verlag Berlin Heidelberg 2012

"i.smap" module, and (b) the classification of land cover polygons "borrowed" from the Spanish land parcel identification system (SIGPAC).

8.1 Introduction

Image classification can be defined, as Schowengerdt (2007) briefly puts it, as "the process [used] to produce thematic maps from imagery." Attempts to carry out such processes on digital data have typically been, since the early days, based on algorithms capable of assigning a label to *each* of the pixels in the image. Usually, this was a reasonable approach to be used with medium to high spatial resolution imagery (e.g. Landsat, 30 m; SPOT 4, 20 m), as the relatively large area covered by each pixel meant that an average radiometric response of the actual land cover on the terrain was captured. A significant number of these well-known techniques are available today, including parametric (e.g. maximum likelihood) and non-parametric (e.g. k-nearest classification) methods, but their effectiveness can become compromised as spatial resolution of the images tend to increase. The reason for this is related to the fact that in very high spatial resolution imagery (e.g. QuickBird images – 2.4 m – digital aerial photography, scanned aerial photography), real world objects tend to appear as groups of pixels of relatively high internal variability due to the heterogeneous spectral response of their surface: when the area covered by each pixel is not large enough to average those responses, traditional pixel-based classifications tend to perform poorly, separating into different classes pixels which are simply different parts of the same cover (e.g. tree crowns in direct sunlight and tree crowns in shadow).

8.2 Object Oriented Classification

High as it may be the internal variability of pixels belonging to the same cover, it is generally lower than that of the whole image (Laliberte et al. 2004), meaning that a certain degree of autocorrelation is usually present. Object-oriented classification methods aim to use this autocorrelation to classify *groups* of pixels (objects) rather than individual pixels. They have demonstrated to be powerful techniques that generally yield better results, in terms of accuracy, than traditional per-pixel approaches (Laliberte et al. 2004; Gitas et al. 2004; Marignani et al. 2008), and their success has been such that its availability has become an essential part of the advertising efforts of many proprietary software packages (e.g. eCognition[1]). A crucial phase within any process of object-oriented classification is the segmentation of the original image into homogeneous groups of pixels. Such methods often include a previous spatial-spectral segmentation process intended to divide the original image in a number of objects, which may then be classified separately. The integration of these two phases (segmentation and classification) varies depending among methods, and according to it these may be classified into: (a) "data driven" and (b) "knowledge driven" methods. In the former, segmentation is conducted based solely on image characteristics, trying to minimize the average heterogeneity of image objects for a given resolution over the whole scene. Once the segmentation has finished it is up to the user to decide whether the objects created do represent or not real-world categories and which of them might be used in the training phase of the classification. Many algorithms available in proprietary software use this approach. In the latter group ("knowledge driven") the algorithm tries to create homogeneous objects that match the characteristics of previously defined training areas. This is the case of the Sequential Maximum a Posteriori Estimation algorithm (SMAP) available in GRASS GIS (Bouman and Shapiro 1992; McCauley and Engel 1995; Neteler and Mitasova 2004). Both approaches have advantages and drawbacks: a great advantage of the first group is that as the segmentation and classification phases are separate stages, different sources of information can be used in each one; on the other hand, the whole process is more straightforward and (usually) user-friendly in the second group.

8.3 Classification of Panchromatic Imagery

Not surprisingly, due to the rapidly increasing availability of multispectral and hyperspectral products that are captured by a wide variety of sensors, the issue of classification of panchromatic imagery has not aroused much interest and the literature on the matter is rather scarce. In spite of this, it is very appealing for researchers of Land Use/Land Cover Change (LUCC), whose work is largely based on the use of historical aerial photographs: black and white aerial photographs often constitute the basis for LUCC studies, as they are usually available for dates as early as the decades of 1920 and 1930 for many areas of the world (Carmel and Kadmon 1998; Hutchinson et al. 2000; Laliberte et al. 2004).

A number of researches, nonetheless, have dealt with this subject and proposed many different approaches: thresholding segmentation (Laliberte

---

1 URL: http://www.ecognition.com

et al. 2004; Hutchinson et al. 2000; Lahav-Ginott et al. 2001), pixel-based techniques (Carmel and Kadmon 1998; Cots Folch et al. 2007), and, obviously, object-oriented methods (Laliberte et al. 2004; Pillai et al. 2005; Marignani et al. 2008). It is interesting to note that a common characteristic of most of these papers is the limited number of classes used, a consequence of the also limited amount of information carried by the image: usually only two classes (Hutchinson et al. 2000; Laliberte et al. 2004; Pillai et al. 2005; Robinson et al. 2008; Agüera et al. 2008) or three at most (Carmel and Kadmon 1998; Kadmon and Harari-Kremer 1999) are identified. In spite of the reduced number of classes, global accuracy values are usually low and range between 60% (Carmel and Kadmon 1998) and 80% (Kadmon and Harari-Kremer 1999). Sometimes higher values are presented (e.g. up to 90% in Pillai et al. 2005), but in these cases the accuracy refers to the identification of objects (e.g. individual trees) and not pixels, which obviously lead to higher figures (an object may be correctly *detected* even if many of the pixels that form part of it are not). Results are sometimes as low as to make the authors consider the need to manually refine the classification (Marignani et al. 2008).

In an attempt to increase the information that is available for classification purposes, many authors have essayed the use of textural information, a measure of the variability of the digital levels among neighbouring pixels. This has proved successful in many cases (Carmel and Kadmon 1998; Cots Folch et al. 2007), but not always (Agüera et al. 2008). Texture is usually measured with the help of the Grey Level Co-occurrence Matrix (GLCM) as defined by Haralick et al. (1973), and may even be useful to improve the classification results of multispectral images as well (Johansen et al. 2007). Other strategies intended to improve the results of classification include radiometric correction of the original images to take into account the effect of the slope and aspect of the terrain (but not always successfully, Carmel and Kadmon 1998), the use of a low-pass filter before classification (Laliberte et al. 2004) or the use of a majority filter on the classified image (Cots Folch et al. 2007). The two latter are intended to reduce the usual "salt and pepper" effect commonly associated to pixel-based classifications.

8.4 Objectives

As it has been already mentioned above in the text, a particular algorithm for object-oriented classification is currently implemented in GRASS GIS (the Sequential Maximum a Posteriori estimation or SMAP). The way the algorithm works (a "data driven" approach) poses some problems for its use on panchromatic imagery: as both the segmentation and classification phases have to be performed on the same group of images, the use of texture measures (e.g. local variance) as ancillary bands tends to cause confusion between some of the land covers having high variance and border areas between different covers. Other interesting problem is the absence of any control over the size and shape of the objects created, an especially important issue in the case of black and white photographs, where the difference between adjacent covers is reduced to subtle differences of grey levels. While these problems might seem enough reason to turn attention to other pieces of software currently available in the market, other alternatives may solve the user's needs – still using exclusively free software – given some conditions are met. This work presents one of such detour approaches: the use of existing vector layers as segments (objects) which may be then classified using statistical techniques.

In summary, this chapter presents an example of classification of historical (black and white) aerial photographs over an area of about 10 km^2 of rural landscape in the Northwest of Spain, into five broadly defined classes: (1) arable land, (2) pastures, (3) pastures with trees, (4) shrubland, (5) forests. Two different approaches are compared: (a) the standard use of the SMAP as implemented in GRASS GIS (the "i.smap" module), and (b) an alternative approach using independently defined land cover polygons (in vector format) "borrowed" from the Spanish Land Parcel Identification System (LPIS), classified with the help the free statistical package R (R Development Team 2008).[2]

8.5 Material and Methods

8.5.1 Material

To compare the performance of both classification techniques, we selected an aerial photography belonging to the so called "B series" photogrammetric flight, flown over Spain by planes and crews of the United States Air Force in the years 1956–1957 (Fernández García 1998). The photograph had been previously scanned to a spatial resolution of 60 cm/pixel and orthorectified to a horizontal accuracy of 4.5 m (root mean square error calculated from independent check points). It covers a traditional agricultural landscape in the province of Lugo (NW of Spain), in which the land covers to be identified (arable land, pastures, shrubland, forest) are

[2] URL: http://www.r-project.org

closely mixed and form a mosaic of patches sized from a few hundred square meters to a few hectares.

On the other hand, the other relevant material used in this work is taken from the Spanish *Sistema de Información Geográfica de Parcelas Agrícolas* (SIGPAC).[3] SIGPAC is the name of the Spanish Land Parcel Identification System (LPIS) that was created in 2003 to comply with regulations of the European Union.[4] The basic cartographic unit in SIGPAC is the cadastral parcel, but a smaller subdivision of it, called *recinto* in Spanish, was created to account for different land uses occurring in the same parcel (Mirón Pérez 2005). The use of *recintos* instead of cadastral parcels to define the segments to be classified was deemed preferable in this work, based on the assumptions that the former would contain more homogeneous land covers in the historic photograph and that the structure of parcels would not have changed significantly in the interim period. Horizontal accuracy of the SIGPAC polygons is estimated to be around 1.25 m (RMSE).

8.5.2 Class Separability and Texture Information

Before starting classification, an initial assessment of separability between classes was conducted. To this end, areas of each pure class (arable land, pastures, shrubs, forest) were manually delineated and a random sample of pixels was extracted from them (5,000 pixels for each class) with the help of the interface between GRASS and R provided by the R package "spgrass6" (Bivand 2008). Shapiro-Wilk normality tests were conducted for the four samples, which showed non-normal distributions in all cases ($p < 0.001$). Separability between classes was shown to be generally poor, as the grey level values of the class "shrubland" largely overlapped with those of "pastures" and "forest" (Fig. 8.1). As a consequence, a great deal of confusion between classes was expected to occur in the resulting classifications.

Because of the existence of such overlaps between the original grey levels of the classes, the use of texture information as ancillary data was believed to have some potential in this case. The main reason for this is related to the fact that the "shrubland" clearly showed a coarser texture than the rest of classes: the big overlap between that class and the rest is due to the existence of pixels with very different grey levels, closely mixed, which is related to a coarse texture. The GRASS module "r.texture" offers

[3] Accessible online at http://sigpac.mapa.es/fega/visor/

[4] Council Regulation (EC) No 1782/2003, Official Journal of the European Union, September 29, 2003.

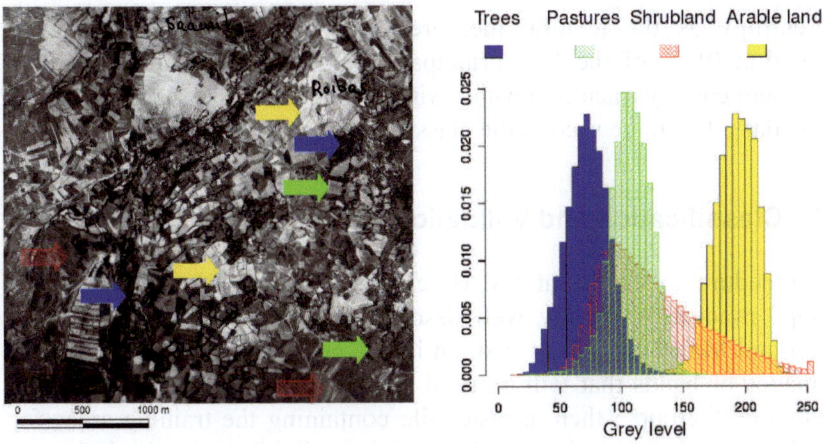

(**a**) Examples on the original image (**b**) Histogram of grey levels

Fig. 8.1. Histogram of grey levels of the four "pure" classes

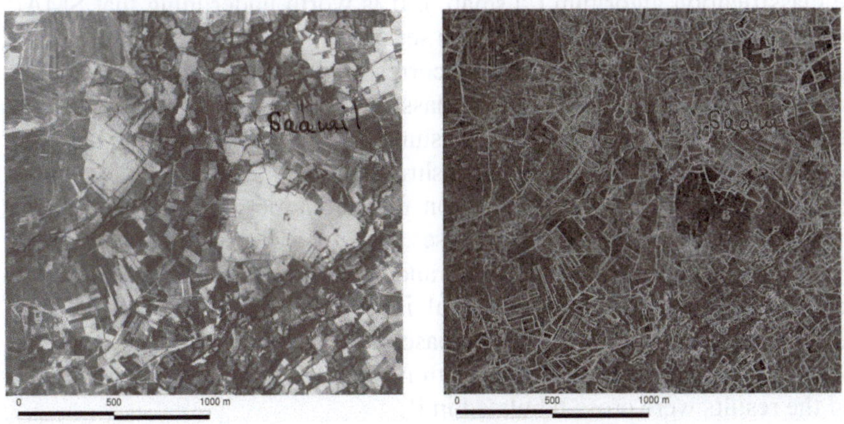

(**a**) Original image (**b**) Sum entropy, window size 5

Fig. 8.2. Texture measure to be used as ancillary data

up to 14 different measures of texture that may be calculated using different window sizes, forming a large array of data from which to choose. Visual inspection of the results obtained with window sizes ranging from 3 to 7 pixels suggested that a window size of 5 could be a good compromise (Fig. 8.2): a smaller window did not generate very different values for the different classes, while larger ones did not only increase the presence of borders – an undesirable effect – but also the processing time. A principal component analysis, done with the "i.pca" module, allowed to identify the sum entropy as the textural measure carrying the largest amount of information (95% of the first principal component): considering all the former, sum entropy calculated with a window size of 5 pixels was selected as the ancillary data to be used in the classification.

8.5.3 Classification and Validation Procedures

The procedure to carry out SMAP classification in GRASS using the "i.smap" module is already well described in the manual pages or by Neteler and Mitasova (2004): first, an image group should be created with the images or bands that will be used in the classification process, using the module "i.group"; then, a raster file containing the training areas for each class may be created using the module "r.digit"; in the third place, a signature file should be generated, using the module "i.gensigset", from the image group and training areas; finally, the last step is simply to call the classification algorithm ("i.smap"). It is worth underlining that SMAP requires the user to manually define a set of training areas which, given the nature of the process, should ideally correspond to homogeneous covers. It is for this reason that the resulting classification would not include one of the classes formerly mentioned (pastures with trees) but only the four "pure" classes: arable land, pastures, shrubland, and forest.

SMAP does not provide any option for automatic validation of results. Thus, for quality assessment purposes, a set of 100 independent check points was randomly generated ("v.random") over the area of study and then manually classified – by visual inspection – into one of the four aforementioned categories. The database of the vector point layer was then updated with the categories resulting from the classification ("v.what.rast"), and the results were cross-tabulated in R.

The alternative classification procedure proposed in this work is a bit less straightforward. As a first step, it requires the vector database to be updated with some univariate statistics taken from the base images (in this case, the original image and the texture image), something that can be accomplished in GRASS using the "v.rast.stats" module. The basic statistics calculated by this module refer (except the number of pixels in each polygon) to the digital levels of the image: minimum, maximum, range, mean, standard deviation, variance, coefficient of variation, sum. Then a sample of polygons has to be selected to create a training set. Given that some of the polygons may contain mixed covers, in this case a mixed category "pastures with trees" was also included in the training sample. Once the vector database contains the information about digital levels enclosed by each polygon and the training sample has been selected, it can be exported to R using either the conventional command "read.dbf" (which can read the GRASS vector database "externally") or using the more advanced R-GRASS interface provided by already mentioned R package "spgrass6". Once the database has been loaded into R, the remaining vector polygons not selected as training may be classified by means of a supervised decision tree. Although the base packaging of R already offers some suitable algorithms (e.g. the "rpart" command), we propose instead the use of the J48 algorithm available in the R package "RWeka" (Hornik et al. 2009), the R implementation of the free data mining software Weka (Witten and Frank 2005). J48 is an implementation of the free algorithm C4.5 by Quinlan (1993). The result is a decision tree like the one in Fig. 8.3.

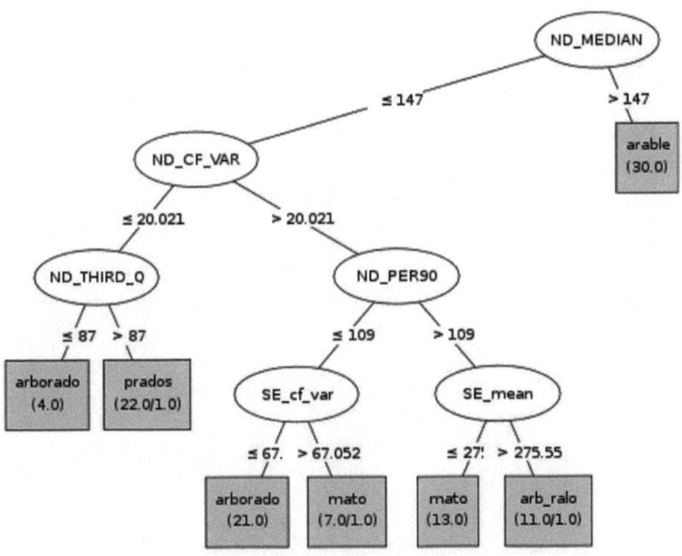

Fig. 8.3. Decision tree generated with J48

The main reason to choose J48 instead of other alternatives implemented in R by default is the possibility that it offers to carry out cross-validation (jackknife) using the training sample. Cross-validation allows the user to avoid the selection of an independent check sample or, in other words, to use all the information available as part of the training set while still having an assessment of classification accuracy coming from an independent sample.

8.6 Results

The results of the two methods compared in this work are radically different, first of all, because the result of SMAP is a raster map while the result of the alternative method is a vector map. Obviously, this translates itself in a different visual aspect (Figs. 8.4 and 8.5) and slightly different levels of usefulness depending on the user's final purposes. The comparison of the accuracy levels corresponding to each method should accordingly be taken with some caution, as neither the subject assessment (classified *pixels* versus classified *polygons*) nor the methodology used (set of independent random points versus cross-validation) are equivalent. Nevertheless, it is clear from the numeric results (Tables 8.1 and 8.2) that the alternative approach suggested in this work shows a much higher global accuracy value (88% against 56%).

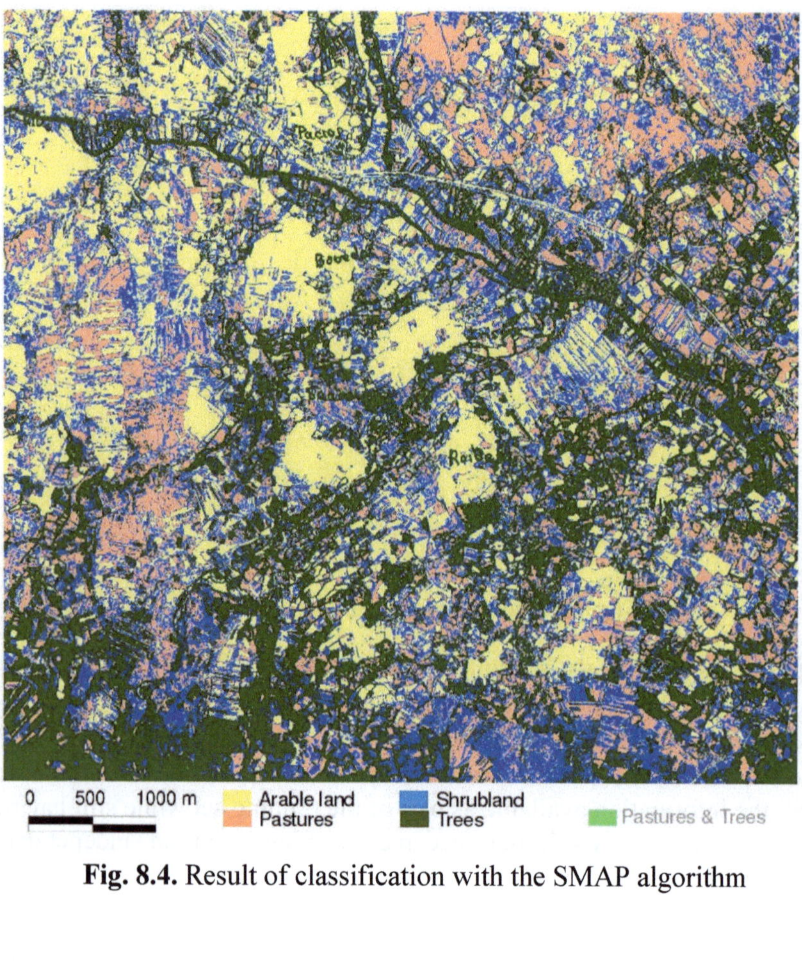

Fig. 8.4. Result of classification with the SMAP algorithm

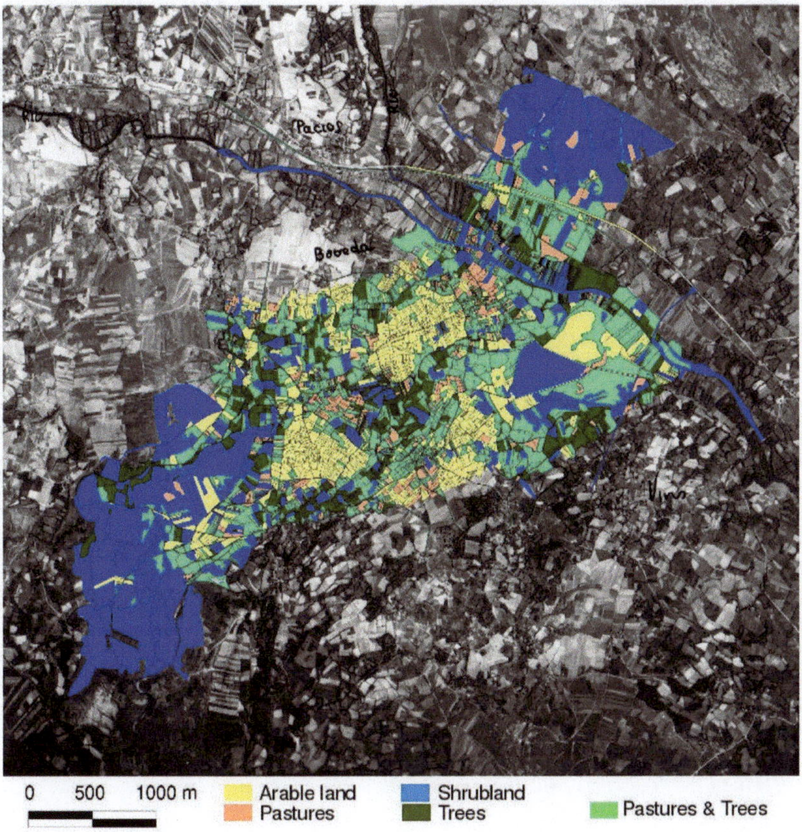

Fig. 8.5. Result of classification of existing vector polygons with J48 algorithm

Table 8.1. Confusion matrix for SMAP classification

Classification	Reference				User's accuracy	Producer's accuracy
	1	2	3	4		
1. Arable	22	1	5	2	73.3%	91.7%
2. Forest	0	13	8	3	54.1%	76.5%
3. Shrubland	2	2	13	12	44.8%	38.2%
4. Pastures	0	1	8	3	25.0%	60.0%

Global accuracy: 56/100 (56%). Assessed using independent points.

Table 8.2. Confusion matrix for the classification with J48

Classification	Reference					User's	Producer's
	1	2	3	4	5	accuracy	accuracy
1. Arable	30	0	0	1	0	96.7%	100.0%
2. Forest	0	23	2	2	0	85.2%	85.2%
3. Shrubland	0	2	16	2	0	80.0%	80.0%
4. Pastures	0	1	0	16	0	94.1%	76.2%
5. Past. w/ trees	0	1	2	0	10	76.9%	100.0%

Global accuracy: 95/108 (88%). Assessed using 10-fold cross-validation.

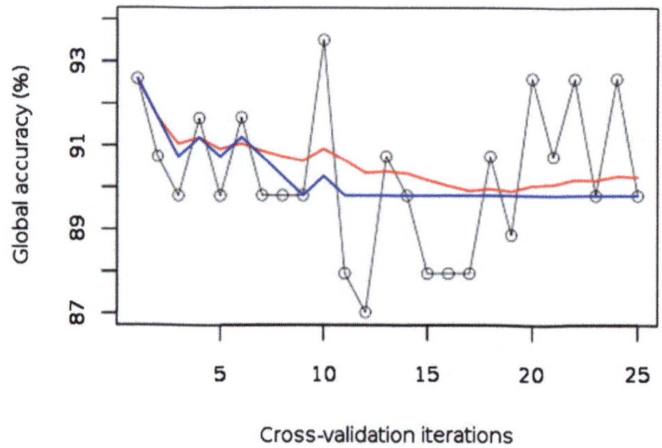

Fig. 8.6. Estimated accuracy of classification with J48 in different iterations

Another issue worth mentioning is the fact that both accuracy assessments presented in the tables result from just *one* independent point sample or cross-validation test. It is not hard to realize, thus, that the resulting values should be taken as an estimate of the real one, and that the user has not any means to guess how close to reality it is. While repeating the assessment for SMAP classification would involve a time-consuming process of selecting new independent points, cross-validation may be repeated several times without additional effort. The results of 25 cross-validation iterations are shown in Fig. 8.1: it appears clearly how the global accuracy estimate can vary considerable among iterations but also how the mean (red line) and median (blue line) tend to stabilize around 90% (Fig. 8.6).

8.7 Discussion and Conclusions

Classification of panchromatic aerial photographs has not been paid much attention by researchers or software developers, reasonably enough as they fulfil a very specific niche in the research community and the possibilities they offer are very limited given the extremely low amount of information that they provide. Nevertheless, a small number of papers exist on the subject, all of them characterized by relatively low results in terms of accuracy. While many of the usual techniques used for multispectral images are also potentially useful – especially when the original panchromatic image is complemented by texture information – it is evident that object-oriented methods are likely to provide the best results. Nevertheless, even in those cases, some of the methods show shortcomings: this is the case of the SMAP algorithm provided in GRASS. In this chapter, an object-oriented approach to classification of panchromatic aerial photographs based on the use of existing vector polygons as image segments is shown, which compared to the original performance of SMAP allows for a significant increase in the accuracy of the final classification. The proposed approach makes use of the possibilities of interaction between the free geographic information system GRASS and the free statistical package R, allowing the whole process to be carried out using exclusively free software. Furthermore, the approach should be possibly used in any country with a good quality public cadastre system, among them all EU countries – in which the existence of a public Land Parcel Identification System is compulsory according to EU regulations. It could also be used to good effect in some phases of the periodic update of the land cover database that these Parcel Systems require, using recent multispectral aerial photographs instead of historic data.

The use of existing vector polygons – be cadastral parcels or subdivisions of them – should be regarded as a simplification of the reality portrayed in historic photographs, as the historic land covers do not necessarily follow the structure of parcels. Moreover, the use of the technique requires not only a good co-register of the vector database and the orthophotograph but also that the structure of parcels would not have changed significantly in the period. On the other hand, the resulting vector database updated with historic land cover greatly facilitates comparison with current land cover included in cadastral databases, an efficient approach to LUCC studies.

References

Agüera F, Aguilar FJ, Aguilar MA (2008) Using texture analysis to improve per-pixel classification of very high resolution images for mapping plastic greenhouses. ISPRS J Photogramm Remote Sens 63(6):635–646

Bivand R (2008) spgrass6: Interface between GRASS 6 geographical information system and R. R package version 0.5-3. URL: http://grass.osgeo.org/

Bouman C, Shapiro M (1992) Multispectral image segmentation using a multiscale image model. In: Proceedings of IEEE international conference on acoustics, speech and signal processing, San Francisco, pp III–565–568

Carmel Y, Kadmon R (1998) Computerized classification of Mediterranean vegetation using panchromatic aerial photographs. J Veg Sci 9:445–454

Cots Folch R, Aitkenhead MJ, Martínez Casanovas JA (2007) Mapping land cover from detailed aerial photography data using textural and neural network analysis. Int J Remote Sens 28(7):1625–1642

R Development Core Team (2008) R: A Language and Environment for Statistical Computing, R Foundation for Statistical Computing, Vienna, Austria. ISBN 3-900051-07-0. URL: http://www.R-project.org

Fernández García F (1998) Las primeras aplicaciones civiles de la fotografía aérea en España. 1: El Catastro y las Confederaciones Hidrográficas. Ería 46:117–130

Gitas I, Mitri G, Ventura G (2004) Object-based image classification for burned area mapping of Creus Cape, Spain, using NOAA-AVHRR imagery. Remote Sens Environ 92(3):409–413

Haralick RM, Shanmugam K, Dinstein I (1973) Textural features for image classification. IEEE Trans Syst, Man Cybern 3(6):610–621

Hornik K, Zeileis A, Hothorn T, Buchta C (2009) RWeka: an R interface to Weka. R package version 0.3-15. URL: http://CRAN.R-project.org/

Hutchinson CF, Unruh JD, Bahre CJ (2000) Land use vs. climate as causes of vegetation change: a study in SE Arizona. Glob Environ Chang 10:47–55

Johansen K, Coops NC, Gergel SE, Stange Y (2007) Application of high spatial resolution satellite imagery for riparian and forest ecosystem classification. Remote Sens Environ 110:29–44

Kadmon R, Harari-Kremer R (1999) Studying long-term vegetation dynamics using digital processing of historical aerial photographs. Remote Sens Environ 68:164–176

Lahav-Ginott S, Kadmon R, Gersani M (2001) Evaluating the viability of Acacia populations in the Negev Desert: a remote sensing approach. Biol Conserv 98(2):127–137

Laliberte AS, Rango A, Havstad KM, Paris JF, Beck RF, McNeely R, González AL (2004) Object-oriented image analysis for mapping shrub encroachment from 1937 to 2003 in southern New Mexico. Remote Sens Environ 93(2):198–210

Marignani M, Rocchini D, Torri D, Chiarucci A, Maccherini S (2008) Planning restoration in a cultural landscape in Italy using an object-based approach and historical analysis. Landsc Urban Plan 84(1):28–37

McCauley JD, Engel BA (1995) Comparison of scene segmentations: SMAP, ECHO and maximum likelyhood. IEE Trans Geosci Remote Sens 33(6):1313–1316

Mirón Pérez J (2005) El catastro y la reforma de la Política Agraria Común de la U.E.: la implantación del SIGPAC. CT: Catastro 54:7–40

Neteler M, Mitasova H (2004) Open source GIS: a GRASS GIS approach, 2nd edn. Springer, New York

Pillai RB, Weisberg PJ, Lingua E (2005) Object-oriented classification of repeat aerial photography for quantifying woodland expansion in central Nevada. In: 20th Biennial workshop on aerial photography, videography, and high resolution digital imagery for resource assessment, Weslaco, 4–6 Oct 2005

Quinlan JR (1993) C4.5: programs for machine learning. Morgan Kaufmann Publishers, San Mateo

Robinson TP, van Klinken RD, Metternicht G (2008) Spatial and temporal rates and patterns of mesquite (Prosopis species) invasion in Western Australia. J Arid Environ 72(3):175–188

Schowengerdt RA (2007) Remote Sensing. Models and methods for image processing, 3rd edn. Academic, Burlington

Witten IH, Frank E (2005) Data mining: practical machine learning tools and techniques, 2nd edn. Morgan Kaufmann, San Francisco

[illegible]

Part 3: Cartography and Standards

Chapter 9 Sensor Web Enablement Standards for Groundwater Monitoring

Tiana Randrianarivelo[1], Pierre Lagarde[1], Vincent Heurteaux[2]

[1]BRGM, 3 avenue Claude-Guillemin, BP 36009, 45060 Orléans, France
{randrianarivelo, p.lagarde} @brgm.fr
[2]Geomatys, 24 rue Pierre Renaudel 13200 Arles, France
vincent.heurteaux@geomatys.com

Abstract. The French Geological Resources and Mining Bureau BRGM (Bureau des Ressources Géologiques et Minières) manages various environmental monitoring sensor networks such as their national piezometer network used to follow groundwater levels. The sensors on these networks have been integrated into the bureau's information system allowing the data to be used for analysis in studies of water consumption, of water quality, and of agricultural use. The Open Geospatial Consortium (OGC) initiates the concept of the Sensor Web Enablement (SWE), which aims to provide standardized web based interface to these sensors and standardized sensor and sensor data models. Using SWE in Web Sensor Networks potentially improves discovery of and accessibility to data through web based search and data exchange mechanisms and potentially reduces costs by standardizing tasking and monitoring interfaces to the devices. In this chapter we describe our software system, the SWE-Piezo system, which extends the existing BRGM data acquisition and dissemination system by adding a standards based, web presence for the groundwater data. The system conforms to the core standards defined by the SWE initiative of the OGC. The SWE-Piezo objectives are to retrieve and access groundwater data. The SWE-Piezo system includes a web client viewer prototype[1] based on Adobe flex technology, for accessing piezometers and piezometers data. Our experience demonstrates a quick approach to bringing

---

1 http://swe.brgm.fr/

E. Bocher and M. Neteler (eds.), *Geospatial Free and Open Source Software in the 21st Century*, Lecture Notes in Geoinformation and Cartography,
DOI 10.1007/978-3-642-10595-1_9, © Springer-Verlag Berlin Heidelberg 2012

the interoperability, scalability, and flexibility of the sensor web to existing sensor data collection networks.

9.1 Introduction

With the effects of the global warming, environmental data have become more and more useful for earth monitoring. Sensor networks are often used to achieve monitoring of physical phenomenon. Information Systems are built around those sensor networks to manage data transmission, data storage and data access.

Managing various sensor networks is a challenge for several reasons:

- It is more practical to have flexible (adaptable) Information System for each sensor networks. It is not necessary to reinvent the wheel each time a new Information System is needed for a new deployed sensor network.
- Decision making usually requires more than one sensor observation or phenomenon observation, thus there is a need of homogenous sensor data access or sensor data format. Interoperability is an ubiquitous need to allow rapid and easy access of heterogeneous data sources.
- In order to fast access (by end-users) the information, enabling the sensor network through the web may be a challenge. Indeed, bringing the sensor to the web in an interoperable way, and advising end-users to rather use an interoperable way may cause some constraints: interoperability, old sensor access, specific data formats or access methods ...

Service based architecture is now emerging in the information technology world. An important issue is to make sensor network available on the web: we call this concept Web Sensor Networks.

In this chapter, we discuss about the collaborating research work of BRGM and GEOMATYS on testing emerging technologies such as the Sensor Web Enablement (SWE) standards for the groundwater monitoring. First we will give a description of the SWE concept and other existing standards. Then we will present the design of our SWE-Piezo system followed by its implementation. In the fourth part of this chapter, we will illustrate the use of SWE trough the SWE-Piezo system. And finally, we will present our ongoing and future work on SWE implementation.

9.2 State of the Art

In order to design the SWE-Piezo system, we have two main criteria: interoperability and scalability. We require interoperable Sensor Based Information System, based on open sensor standards. The system should be scalable and extensible. Scalability may be an ambiguous term. Duboc et al. (2007) did an interesting review of scalability definition, underlying the impact of varying system variables to qualify the scalability of a system. We require scalable sensor based information system, which scaling variables are the sensor type and the sensor domain application. In this section, we present mainly the Sensor Web Enablement (SWE) initiative of the Open Geospatial Consortium (OGC), and some other interesting standards dealing with sensors and water domain.

9.2.1 The Sensor Web Enablement

The SWE (Botts et al. 2007; Ingo 2008) initiative of the OGC[2] aims to develop and publish standards for the services, interfaces, and data exchange patterns within information systems built around sensors. The OGC is a non-profit, international collaboration between government agencies, private enterprises, and research institutions which develop standards for geospatial services through proposal based discussion and consensus agreement. The OGC aims to promote interoperability in Geographic Information System (GIS). The SWE initiative comprises one of the efforts of the OGC. The BRGM has been actively participating in the OGC in general and in the SWE effort in particular. The SWE specifications aim to enable sensor and sensor data being described, discovered, accessed and tasked through the web in a standard manner, facilitating interoperability. These standards can be split to two main categories.

The first set of standards deals with sensor description and data definition:

- The Sensor Model Language (SensorML)[3] provides a descriptive model for sensors and an XML schema for describing the modeled sensors. The sensor models can describe actual physical sensors or describe abstract sensors such as a process; the SensorML schema can be used to describe sensor metadata such as sensor characteristics, locations, and properties.

---

2 http://www.opengeospatial.org/

3 http://www.ogcnetwork.net/SensorML

- The Transducer Model Language (TransducerML) provides a conceptual approach and XML Schema for supporting real-time streaming of data to and from sensor systems, describing the system as well as the data encoding.
- The Observation and Measurements (O&M) provides a conceptual model and XML Schema for encoding observations and measurements from a sensor, along with the encoding usable for real-time or archived data.

The second set of standards relates to services:

- The Sensor Observation Service (SOS) provides a standard web service interface for accessing, filtering and registering or inserting sensors and observations.
- The Sensor Alert Service (SAS) provides a standard web service interface for managing alerts using a 'publish and subscribe' approach.
- The Web Notification Service (WNS) provides a standard web service interface for asynchronous delivery of messages or alerts from any other web service.
- The Sensor Planning Service (SPS) provides a standard web service interface for tasking sensors, or requesting user-driven acquisitions and observations.

Among SWE implementations, 52° North's SWE implementation is the reference implementation for the OGC's compliance and interoperability testing initiative (CITE).[4] The SWE initiative has proven its objectives through its use within some European projects like SANY, OSIRIS and GEOSS.

9.2.2 Other Sensor Standards

Among other available sensor standards, there is the IEEE P1451.[5] It is a family of Smart Transducer Interface Standards. It describes a set of open, common, network-independent communication interfaces for connecting transducers (sensors or actuators) to microprocessors, instrumentation systems, and control/field networks. This family has for instance standards for commands and operations, and standards for an object model describing the behaviour of smart transducers. The key feature of these standards is the definition of Transducer Electronic Data Sheets (TEDS) which describes a standardized transducer model.

---

[4] http://cite.opengeospatial.org/reference

[5] http://ieee1451.nist.gov/

Groundwater monitoring is our main use case for the SWE-Piezo system. Various water standards are currently available. Among these standards, there is the French water specification SANDRE[6] and WaterML[7]: both are advanced specifications of water model and water data queries methods. There is also the GroundWater Markup Language[8] initiative, a GML (Geography Markup Language) extension to exchange groundwater information. It is really a specific domain oriented standard suitable for our SWE-Piezo system. The OGC is actually very aware of such standards. There are now promising efforts of OGC working groups initiative[9] in the water domain (like the WaterML 2 development) in order to make advanced standardization and harmonization of various existing standards.

The SWE initiative offers open standards. It is supported by an increasing community of private, public and academic institutions. SWE data models and services interfaces offer high level sensor data abstraction, making them generic to support various sensor based information system requirements. TEDS is not an open standard, is designed for transducers and actuators (physical devices), and is more suitable to manufacturers. An interesting comparison between SensorML and TEDS is proposed at (Peizhao Hu et al. 2007). It proposes a combination use of TEDS and SensorML. Indeed, TEDS actually is efficient in the sensor hardware level and SensorML is more suitable to the sensor application level.

Water related standards are suitable to our SWE-Piezo system, they offer advanced works on water data modeling and water data access specifications. However, we intend to extend the SWE-Piezo system to other domains (weather data, earth-quake data, landslide models …). The scalability of the system may be lost by designing water standards based sensor information system.

So, the SWE initiative fulfils our goals in order to design an interoperable, scalable and flexible sensor based information system. We mainly focus on the use of SWE. We do not focus on sensor networks architecture, sensor data storage management or a global sensor based information system architecture. In our future work, we will be investigating some interesting advanced works such as the distributed GIS proposed at (Liang et al. 2005) with the concept of Dynamic Hash Table, the indexing of real-time spatio-temporal data proposed at (Laurini et al. 2005) and the results of the European project SANY, the sensor architecture (Klopfer and Simonis 2009).

---

6 http://sandre.eaufrance.fr/

7 http://his.cuahsi.org/wofws.html

8 http://ngwd-bdnes.cits.rncan.gc.ca/service/api_ngwds/en/gwml.html

9 http://external.opengis.org/twiki_public/bin/view/HydrologyDWG/WebHome

9.3 Design

We developed a prototype system applying the SWE specifications to a groundwater monitoring network. We intend to evaluate the applicability of the SWE standards to a system geared towards providing end-users with access to near real-time data using an open and standardized interface.

9.3.1 Groundwater Existing Information System

The BRGM groundwater monitoring sensors are typically piezometers, and are deployed in several networks all over the French territory. An overview of the groundwater acquisition system is illustrated in Fig. 9.1 Data from sensors networks are stored locally (the territory is subdivided in regions) before being sent and archived in a centralized database. The public access of data is then provided by the ADES web portal.[10]

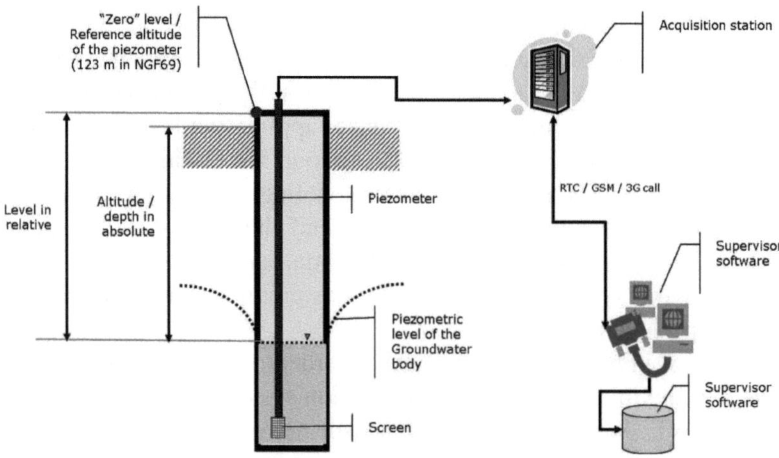

Fig. 9.1. Sensor system acquisition

Our work was constrained by characteristics of the current data acquisition system. The acquisition system uses proprietary software and data formats to manage sensor device requests and responses (our acquisition system is provided by OTT[11]). Data are only available through this software preventing us from communicating directly with the sensors. The acquisition system software's role is mainly to assure sensor request operation and sensor data

[10] http://www.ades.eaufrance.fr/

[11] http://www.ottfrance.com/

storage. To limit energy use, sensors are only tasked daily or weekly. Groundwater data are not published immediately to the ADES system but undergo a validation process to verify raw data. The full data sets are stored locally in each region, with only selected observations being sent to the centralized ADES system. We therefore integrated our new information system within these constraints.

9.3.2 Groundwater Information System with SWE Implementation

The architecture we used for our new information system is presented in Fig. 9.2 which focuses on the SWE components of the implementation. We use the SensorML schema to describe our piezometers and the O&M schema to encode groundwater data. We implemented a SOS server and client, using the client to send data observations in batch transactions and the SOS server to record the transactions and provide web access to the recorded observations.

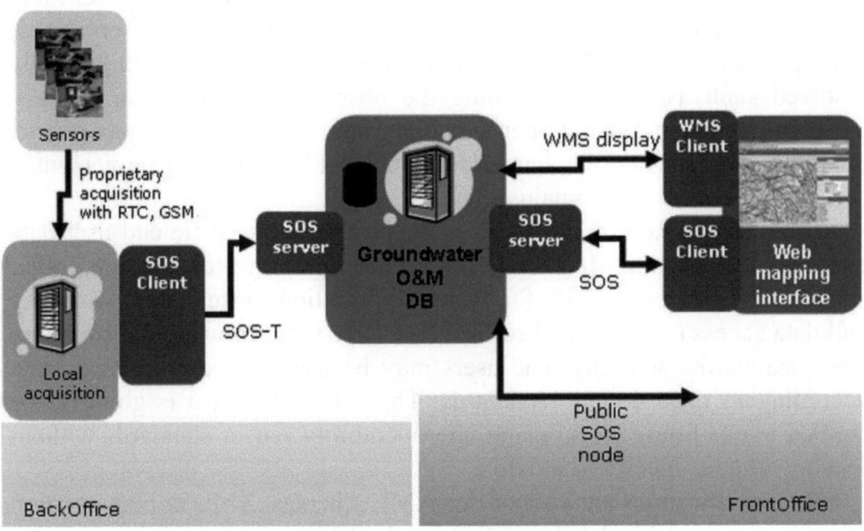

Fig. 9.2. System architecture

The new information system architecture has three main layers:

- The server layer contains a SOS server and a Database Management System (DBMS) with a schema implementing O&M. The SOS server is a Web Service used for requesting sensor or observation data (non transactional operations) and for populating the O&M DB with sensor or observation data (transactional operations).

- The back-office layer which automatically publishes sensor data to the server layer. As mentioned previously, our action scope is prevented from direct acquisition by the proprietary software system so we acquire data already obtained from the sensor. We use the SOS transactional operations (SOS-T), implemented as a SOS Client, to enable publication of sensor data.
- The front-office layer provides a public facade for the data through the SOS server which handles requests for sensor or observation data. This part of the system would typically be open for processing applications. This layer was implemented following the SOS standard to enable SOS server requesting and response analysis.

9.3.3 Experience Learned and Remarks

Open sensor standards provide one way to enable web based access to sensor systems and interoperability. The SWE data encoding (O&M and SensorML) allowed us to describe our data (piezometers or observations) without major issues. However, there were some minor points not fully resolved such as issues describing the observations validation process. SWE specifications enable interoperability but using standards may restrict data description when the standard has not yet been widely implemented (various end users and domains).

Sensor data generally use their own data format (specific end user data encoding for example). Moreover, in our groundwater domain, other specifications already exist for water information system data encoding and data access (as presented earlier). We have to be careful on using these SWE standards. Actually, end-users may be used to exploit sensor data according to their domain standards. This particular case is going to be studied in our future work: using interoperability sensor standards without loosing domain specific standards.

SWE specifications are a set of rich XML schemas. XML implementation offers human readable data but unfortunately, XML messages and documents cost system resources: mainly xml processing and xml message transport. A balance between standardization and system performance should be explored. As instance, sending REST requests are lightweight than using Post XML requests (REST requests are not actually standardized requests expect the GetCapabilities operation).

9.4 Implementation

Our new information system included five components, a SOS server, SOS clients, a Web Mapping Service (WMS) server, a WMS client, and a web application. All transactions between clients and servers use the standard http protocol.

9.4.1 Server Layer

The two servers, the SOS server and WMS server, were developed in the framework of the Constellation-SDI project.[12] Constellation-SDI is developing a series of OGC standard web services, in this case only a SOS and a WMS, on top of a server engine for data storage, retrieval, manipulation, indexing and search. These servers are standard Java EE applications built using both standard web services libraries such as JAX-B, JAX-WS, and newer libraries such as JAX-RS. The servers can therefore run in any servlet container such as Glassfish, Tomcat, Jboss, or Grizzly.

The web service layer can handle any communication protocol by providing a protocol specific entry point for each supported protocol, notably the HTTP based "REST", "SOAP" and "RESTful" protocols, all delegating to a common backend worker which performs the actual logic of each service request. The communication pattern for these services, including the message contents and exchange patterns, has been standardized by the OGC to ensure interoperability of the server with any client built to handle the standard.

9.4.2 Data Storage Management

Underneath this communication layer, the server engine provides data access, manipulation and search. The SOS server relies on a generic meta-schema for Observation and Measurement databases built with the Post-GRID (part of Constellation) data access, manipulation, indexing and storage system. Post-GRID provides a common Java facade to a database storage system which allows the manipulation of the database and a tightly coupled input/output system which can generate new entries in the database from certain Java objects and can generate new Java objects from queries made against the database. This goes far beyond common persistence approaches because the generated objects can be entirely new, based on

[12] http://constellation.codehaus.org/

the result of some complex query, rather than merely being the retrieval of a new instance identical to the object originally stored. We used the Postgresql DBMS as the underlying database manager to store our O&M data. For its part, the WMS server uses a combination of geographic data sources stored on file and prepared as imagery on demand.

9.4.3 Sensor Data Publishing and Accessing

A SOS client was built to provision the SOS server by periodically sending new values. The client runs as a daemon service on the server of the traditional information system which aggregates sensor data. Periodically this daemon reviews new data arrivals, transforms these new entries into standard SOS transaction messages, and sends these messages to the SOS server.

To illustrate the ability of standard WMS,WFS and SOS clients to access and use the data on the web server, we integrated a mapping and querying component into a web portal[13] (as illustrated in the following Fig. 9.3 an Adobe flex based client is used as host web client for sensor data monitoring). This new component combined the ability to act as a client to the WMS to obtain the base maps and a module to act as a client to the SOS to obtain the sensor observation. These clients were combined with a graphical mapping interface to display the values and their spatial location.

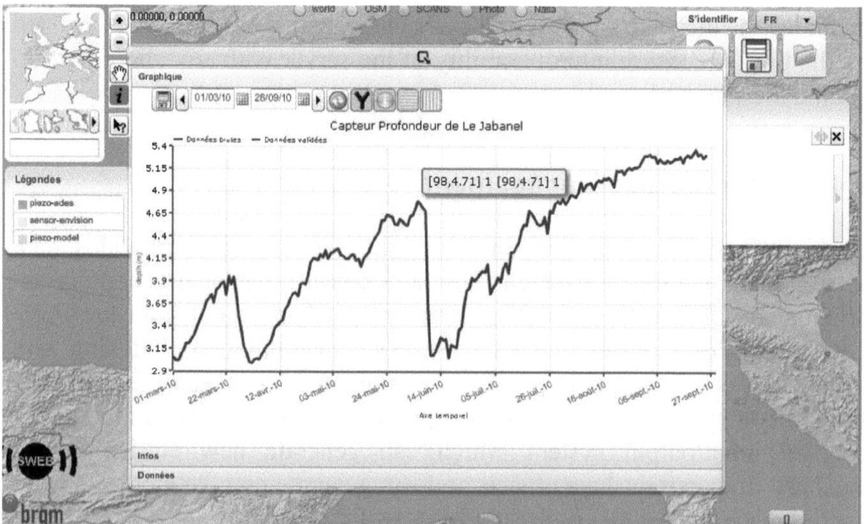

Fig. 9.3. Client illustration

[13] http://swe.brgm.fr/

In order to simplify sensor discovery, we have stored Sensor Metadata in a Relational Database Management System based on a meta-model. This meta-model was inherited from the MDweb project,[14] and chosen for its ability to support all kinds of normalized metadata. In that case, sensor metadata are represented using the SensorML OGC specification which has been described in a standardized dictionary and incorporated in the database.

9.5 Use of SWE Standards

Within our SWE implementation, we are using SensorML 1.0.1, O&M 1.0.0 and SOS 1.0.0.

9.5.1 SensorML: for Sensor Encoding

SensorML can, but generally does not, provide a detailed description of the hardware design of a sensor. Rather it is a general schema for describing functional models of the sensor (Botts and Robin 2007). We are using the following information to describe our piezometers:

- **Sensor type**: sensor type may be a SensorML system or a SensorML component. A SensorML component is actually one and only one physical sensor. A SensorML system is composed of one or more SensorML components. All of our piezometers are SensorML Systems. Indeed, some of our piezometers have more than one sensor device (piezometer device, temperature device or conductivity device)
- **Keywords**: GroundWater, Piezometer, Geosciences
- **Identification**: sensor metadata like sensor name, manufacturer, producer, device information
- **Classification**: water body, monitoring network, administration area …
- **Constraints and contacts**: rights and organization / roles
- **Characteristics:** max depth of the well where the piezometer is located
- **Location:** location of the sensor
- **Interfaces**: exchanges protocols
- **Inputs and outputs**: water level as input and water depth (temperature and conductivity) as output(s)
- **Components and connections:** for SensorML systems only, components list and connections between the components of a system.

[14] http://www.mdweb-project.org/

SensorML provides a generic and sufficient description of our piezometers

9.5.2 O&M: for Sensor Data Encoding

The complete O&M specification is available at Cox (2007) . An observation contains basically:

- **Feature of interest**: observation related feature, in our case, we have the well as feature, represented by a sampling point
- **Sampling time**: time period of the proposed observation (time period or time instant)
- **Observed property**: phenomenon which is (GroundWater) depth.
- **Procedure**: observation source, the piezometer
- **Result**: observation data: observation encoding and observation values. Observation Result can be any type of data (including domain specific data type). We are using SWE DataArray to encode our observations

A SWE DataArray contains:

- **Values metadata**: number of values, fields declarations (Time and depth), values encoding (we use text blocks to render our data)
- **Values**: well formed sensor data according to the values metadata

Examples:
Text block encoding:
<swe:TextBlock tokenSeparator="," decimalSeparator="." block Separator="@@"/>
Raw values:
2009-04-03T12:00:00,8.35@@2009-05-22T12:00:00,7.46@@
2009-07-17T12:00:00,8.29@@

We have three blocks separated by "@@", each block has two fields separated by ",". The first field is a Time value and the second field is the (GroundWater) depth value.

GroundWaterML, WaterML or SANDRE data formats can be used to encode the Observation results, as results type can be any type of data. Thus, it makes the O&M flexible.

9.5.3 SOS: for Registering and Requesting Sensor Data

The complete SOS description is available at (Na and Priest 2007). The SOS specification includes various web services interfaces to: request the SOS capabilities, register sensors, insert observations and request sensor description or sensor data.

GetCapabilities

GetCapabilities operation is used to request the SOS server capabilities. The SOS capabilities document contains useful information like:

- **Operation metadata**: available SOS operations with the associated method calls and parameters
- **Offerings list**: an offering is a group of Observations. It provides a mechanism for factoring groups of related observations within a single service instance (Na and Priest 2007). Each offering has a list of associated procedures (piezometers), feature of interests (feature representing the well where a piezometer is located), valid time (time period of available sensor data)

RegisterSensor

This is an optional operation. It is implemented in the SWE-Piezo system, allowing us to register our piezometers. To register a new sensor, a SensorML document along with an Observation template of this sensor is necessary. The Observation Template is an observation sample of this sensor data, and is being used later by the InsertObservation operation.

InsertObservation

This is not a mandatory operation. We implemented it in the SWE-Piezo system to feed the SWE-Piezo system with newly available sensor observations. It basically uses the Observation Template registered with the RegisterSensor, to encode the newly available Observations. The SOS client in the back office part (Fig. 9.2) uses mainly this SOS operation to feed the SWE-Piezo system with new Observations.

DescribeSensor

The DescribeSensor operation allows us to request description of sensors (procedure). A basic request must indicate at least one procedure. We describe our piezometers as SensorML documents; so a DescribeSensor response is only available in SensorML format.

GetObservation

A GetObservation operation requires mainly one or more observation offerings, one or more observedProperty (phenomenon) and a response Format. In the SWE-Piezo system, observation offerings are based on the sensor monitoring network classification. Our piezometers monitoring network are based on administration area or water body. ObservedProperty can be depth, temperature or conductivity. Proposed responseFormat is O&M. We can add one or more procedures, spatial or temporal constraints to the request in order to query specific observations. A GetObservation response is an ObservationCollection, composed of Observation results for each procedure. Procedures are proposed according to the requested offerings and requested constraints.

9.6 Ongoing Work and Future Initiative

We have two main research subjects on using the SWE specification for various purposes: working on sensor web architecture for heterogeneous sensor networks and integrating virtual sensors: non physical sensors. These subjects will allow us to upgrade the SWE-Piezo system the scalability and flexibility of the SWE-Piezo system

9.6.1 Sensor Web Architecture for Heterogeneous Sensor Networks

At the BRGM, we need to do a deeper research work on Information System Architecture around the web sensor networks. Concretely, robust, flexible, reliable and secure information system architecture is needed for managing heterogeneous sensor networks. Heterogeneous sensor networks are wide: sensor networks differ in acquisition system, data encoding, sensor type, communication protocol used, sensor aggregation, observed phenomenon … We assume that SWE specifications could not be enough for having interoperable system before dealing with the heterogeneous aspect of sensor networks.

9.6.2 Sensor Web Enablement and Virtual Sensors

In heterogeneous sensor networks, sensor may be non physical device. We are also interested in the concept of virtual sensor. Virtual sensors are not physical sensors; they are mathematical models process able to simulate sensor observations. Indeed, environment phenomenon is often associated with mathematical models. So it allows observation predictions. SWE specifications also cover non physical sensor. But our research work leads to more issues dealing with heterogeneous information sources used by the mathematical models (models algorithms, models factors, models parameters from other physical or non physical sensors). Our challenge is to provide interoperable Sensor Based Information System with real sensors and virtual sensors from heterogeneous sources, while trying to use existing sensor based information systems at their best (legacy systems should not to be rebuild, they have to be re-used or to be integrated).

9.7 Conclusion

Sensors are anywhere. Various sensor based information system are used to monitor our environment. For a specific organization, scientists, decision makers or public users, accessing multiple sensor data is very useful. Designing interoperable, scalable and flexible sensor based information system and making sensor data available to the internet are necessary to provide common sensor data, deploy new sensor information system, integrate other sensor based information system and fast access sensor data.

The SWE-Piezo system is the BRGM's first step to deal with these requirements, by using the Open Geospatial Consortium (OGC) Sensor Web Enablement (SWE) and SWE-enabling our existing sensor acquisition system. The SWE-piezo system uses some recent technologies like Java based API JAXB and JAX-WS for xml schemas and web service management. We use a new database module access API, Geomatys's PostGRID, able to perform completely new object from some complex query.

The OGC SWE initiative is a promising sensor standard for enabling interoperability. Its abstract data encoding and high level service interface allow SWE specifications to be applied in specific sensor network (use case). The Sensor Model Language (SensorML) standard offers sensor model and encoding; we are able to describe rightly our piezometers without major blocking issues. The Observation and measurements (O&M) standard provides an abstract sensor data model, which suits to our groundwater data. The Sensor Observation Service (SOS) standard offers web based high level interface for accessing sensor and sensor data. We find this standard convenient for responding to our business needs: with its operations like register/describeSensor, or insert/getObservation, piezometers and groundwater data registering and access can be performed and this in a standardized way.

However, the OGC SWE is yet an emerging initiative; adoption of such new technology raises potential issues. Enabling interoperability on information system architecture around sensor networks is constrained by the use of standards, and the balance between new technology and existing ones. While designing interoperable information system, existing sensor based information systems and existing domain specific standards should not be ignored.

In this chapter, we demonstrated the use of OGC SWE to perform groundwater monitoring. Groundwater data can be accessed by web services and in a standardized way. Thanks to the high level sensor data abstraction and service interfaces provided by SWE, the SWE-Piezo system can be re-used to deploy new sensor based information system.

References

Botts M et al (2007) OGC sensor web enablement: overview and high level architecture. OGC 07–165. http://portal.opengeospatial.org/files/?artifact_id=25562. Accessed May 2009

Botts M, Robin A (2007) Sensor model language (v.1.0.0). OGC 7–000, see http://portal.opengeospatial.org/files/index.php?artifact_id=21273. Accessed May 2009

Cox S (2007) Observations and measurements – part 1 – observation schema (v.1.0). OGC 7-022r1. http://portal.opengeospatial.org/files/index.php?artifact_id=22466. Accessed May 2009

Duboc L, Rosenblum D, Wicks T (2007) A framework for characterization and analysis of software system scalability. In : Proceedings of the 6th joint meeting of the European software engineering conference and the ACM SIGSOFT symposium on the foundations of software engineering, Dubrovnik, 03–07 Sept 2007, ESEC-FSE '07. ACM, New York, pp 375–384. http://doi.acm.org/10.1145/1287624.1287679. Accessed May 2009

Ingo S (2008) OGC sensor web enablement architecture, OGC 06-021r2. http://portal.opengeospatial.org/files/?artifact_id=27775. Accessed May 2009

Klopfer M, Simonis I (2009) SANY – an open service architecture for sensor networks. ISBN 978-3-00-028571-4

Laurini R, Servigne S, Noel G (2005) Soft real-time GIS for disaster monitoring. In: Oosterom P, Zlatanova S, Fendel E (eds) Geo-information for disaster management, The Netherlands. Springer, New York, pp 465–480.

Liang SH, Croitoru A, and Tao CV (2005) A distributed geospatial infrastructure for sensor web. In: Comput. Geosci. 31, 2 (Mar. 2005), 221–231. doi:10.1016/j.cageo.2004.06.014. Accessed May 2009

Na A, Priest M (2007) Sensor observation service (v.1.0.0), OGC 06-009r6. https://portal.opengeospatial.org/files/?artifact_id=26667. Accessed May 2009

Peizhao Hu et al (2007) Sensor standards: overview and experiences, In: 3rd international conference on intelligent sensors, sensor networks and information, ISSNIP, pp 485–490, 3–6 Dec 2007. doi: 10.1109/ISSNIP.2007.4496891

Chapter 10 Towards an Automated Label Placement Service

Olivier Ertz[1], Maxence Laurent[1], Daniel Rappo[1] and Eric Taillard[2]

[1]IICT, HEIG-VD, University of Applied Sciences Western Switzerland, Centre Saint-Roch, Avenue des sports 20, CH-1401 Yverdon-les-Bains, Switzerland, {olivier.ertz, daniel.rappo} @heig-vd.ch
[2]MIS, HEIG-VD, University of Applied Sciences Western Switzerland, Route de Cheseaux 1, Case postale, CH-1401 Yverdon-les-Bains, Switzerland, eric.taillard@heig-vd.ch

Abstract. The main focus of this chapter is to introduce a new way to ease the integration of labelling placement functionalities from a given library: a programming-language independent solution based on a web service wrapper. Considering the GeoWeb allowing to combine geodata, mapping, and geoprocessing services to fulfil various needs, it is to give an answer to a current lack: there is no such an approach for delivering advanced labelling functionalities through a standardized web service. The chapter describes a solution to wrap PAL library using standards from Open Geospatial Consortium like Web Processing Service and Symbology Encoding.

10.1 Introduction

While the concept of Digital Earth is becoming reality through the nascent GeoWeb, democratization of GIS and cartography uses are on the way with more and more map makers and map consumers looking for answers to their geographic and cartographic needs. The web is the ultimate platform to discover, explore, share, and even create new content by combination of services. Geospatial data are now widely available through web services and are the base material to create cartographic products

E. Bocher and M. Neteler (eds.), *Geospatial Free and Open Source Software in the 21st Century*, Lecture Notes in Geoinformation and Cartography,
DOI 10.1007/978-3-642-10595-1_10, © Springer-Verlag Berlin Heidelberg 2012

useful for stakeholders. GeoWeb becomes a new paradigm for GIS and cartography.

In this context of map production, placement of labels is an important requirement for maps of quality. To summarize the purpose, it is to label features by a text that should be placed close to them. For instance, in the case of point-features labelling, texts are placed around features and not directly above them, to avoid obscuration of the features. The goal for point-feature labelling is to find a position for each label in such a way that no label overlaps another one or overlaps the symbol which it is marking the point. Finding a good solution of label placements on a map can be hard.

There are many algorithms to find label placement, the reader is referred to (Wolff 2010) for an extensive bibliography. Some have a greedy – i.e. sequential – approach, which means that the first feature of the layer has always more chance to appear than the last one. Others algorithms are clever: they compute the labelling solution in two steps in order to give the same chance to each label to appear. They first predefine arrangements according to features and user preferences. Arrangements are either discrete – a limited number of fixed positions are generated – or continuous – an area in which the label can slide is then defined. In a second step, an effective place for each label is selected with the help of a combinational approach.

Very often, finding a conflict-free labelling is not possible, but the objective is to compute a labelling solution containing as many labels as possible without conflicts. As soon as more than one position is available for placing the label of each object, the problem of finding an adequate solution is NP-hard.

Nowadays, it is rare to find ready-to-use libraries dedicated to label placement. PAL (French acronym for "Automated Label Placement") is an open source project offering such a library with the powerful benefit of a combinational approach. It eases the integration within existing applications that need label placement functionalities. PAL is natively for "C tribes" but offers a Java wrapper. Obviously, other wrappers could be written for more languages.

Nonetheless, the main focus of this chapter is to introduce a new way to ease the integration of labelling placement functionalities from a given library: a programming-language independent solution based on a web service wrapper. Considering the GeoWeb allowing to combine geodata, mapping, and geoprocessing services to fulfil various needs, it is to give an answer to a current lack: there is no such an approach for delivering advanced labelling functionalities through a standardized web service.

With no doubt, the essential standards like HTTP, HTML and others from W3C helped to the success of the omnipresent web. So it is for the GeoWeb, based on Open Geospatial Consortium (OGC) standards, like Web Feature Service to provide GML encoded geodata (Cox et al. 2004), Web Map Service and Feature Portrayal Service (OGC 2007) as cartographic rendering engine and Symbology Encoding (OGC 2006) for the portrayal description. Use of standards helps to favour interoperability from geodata to cartographic contents, so must it be for such a labelling service. This chapter describes a method on how to wrap it with OGC conformity. It follows the idea of a GeoWeb built up with services based on OGC standards which such a labelling service would be a brick of.

10.2 PAL Library, a Study Case

The PAL library is used to describe the method, so is this section about how it works. As the wished labelling web service has to wrap the library, it is first important to know what functionalities it offers and what the inputs/outputs are. Also, a sub-section explains what optimization approach is used to find good placements of labels either for points, lines or polygons.

10.2.1 The Labelling Process

Designed for multi-layers labelling, the PAL library handles layers of points, lines or polygons. Each layer has properties that impact the labelling process:

- Scale range: for which the layer will be labelled
- Priority: to decide which of two conflicting labels from different layers to display
- Obstacles: to avoid labels to be displayed above other features
- Activity status: "Is the layer currently displayed?"
- Labelling status: "Will the layer be labelled?"
- Labelling arrangement: according to the type of feature (free, horizontal, line, centroid …)

In practical labelling, label overlapping is generally forbidden, and there are preferences among the potential positions of a label for a given object. PAL library considers the discrete model, takes the preferences set by the application into account and try to maximize the number of labels displayed without overlap.

10.2.2 Generation of Label Candidate Positions with PAL

Point feature: on a map, a point feature is rendered by a symbol (circle, star, square, etc.) that can be included in a circle of radius *r*. The label of this object shall not overlap with this circle. The candidate positions for a point feature are spread as regularly as possible around this circle (Fig. 10.1). Point features are almost always labelled with a horizontal text in practice.

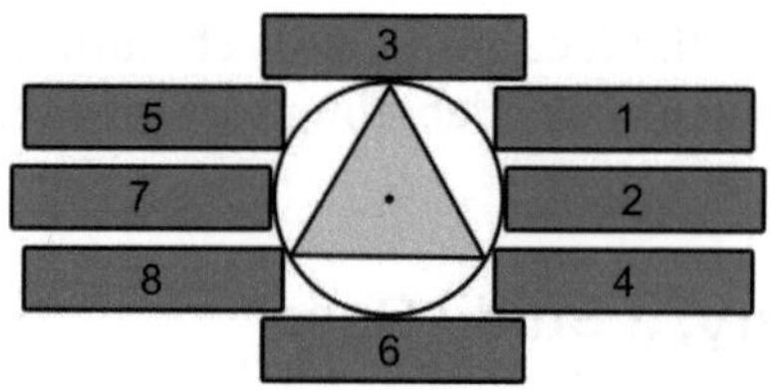

Fig. 10.1. Circle of radius *r* including the symbol and eight candidate positions for labelling this object

Line feature: it is composed of line sections. In practice, the label associated with a line is almost always oriented in a direction locally parallel to the line. Since we only consider rectangular labels, the orientation of the label is determined as follows: let *w* be the width of the label, A and B two points of the line at a distance of *w* each others (the distance being measured along the broken line).

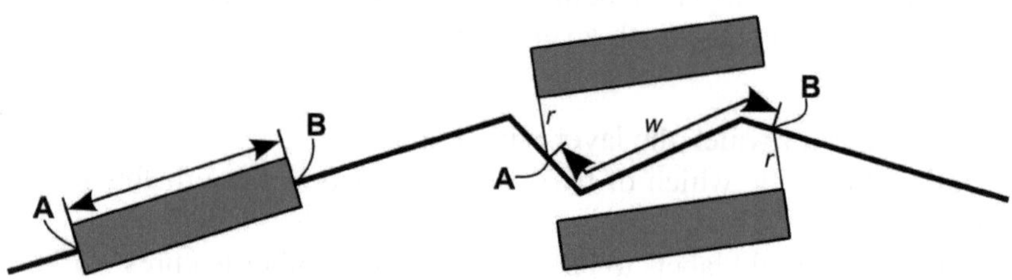

Fig. 10.2. Possible labelling of a broken line: the label is placed on the line or on one side or the other at distance *r*

As shown in Fig. 10.2, the orientation of the label is parallel to the AB segment. The user can specify two ways of labelling a line: either the label is placed on the line – and the AB segment is right on the centre of the label – or the label is on the side or the other of the line – the AB segment is at distance *r* of the label.

Polygon feature: it is the most complex situation and there are different ways to label a polygon. First, a (small) polygon can be considered as a point feature. The candidate generation is then done as for a point feature in that case. Second, the label can be placed on the border of the polygon. The polygon is therefore considered as a broken line and the user can specify the both ways for labelling line-features. Finally the user can specify that the candidate's positions must be inside the polygon (in case the last is sufficiently large; otherwise, the label is only partly inside). For positions that are inside a polygon, the orientation of the label can be specified to be either horizontal or freely chosen by the system. In that case, the polygon is decomposed into a set of (almost) convex polygons. Each of these convex polygons is embedded into a rectangular box of minimal surface. The orientation of the label is parallel to the border of this box. Figure 10.3 illustrates how candidate positions are generated for a non-convex polygon.

10.2.3 Candidate Positions Filtering and Optimizations

Since candidates have been generated for each object to label, a conflict-graph can be computed. In this graph, each node represent a candidate and an edge links two candidates which cannot been displayed together, cause of conflict. The solution of the labelling problem is a maximal stable set of this conflict-graph. Since algorithms designed to find a labelling solution in this graph are very sensitive to the degree of the graph nodes – i.e. average number of edges – the number of candidate positions for each object must be limited. So, only candidates having potential to lead to good solution are preserved. A first filter stage discourages the use of candidates locating over obstacles. A second filter stage selects for each object the p candidates having the lowest cost. A third filter stage, in the spirit of rule $L1$ (Wagner et al. 2001), removes all candidates of a given object that are worse than a candidate of this object having no overlap with candidates of any other objects. This filter stage is recursively applied until no improvements can be obtained.

The Fig. 10.3 illustrates the whole process of generation applied to a polygon with an obstacle. First, the polygon is decomposed into a set of convex ones which are embedded into rectangular boxes. Candidates are generated by covering each box regularly. Candidates' orientation is parallel to the border of the box which contains the candidate. Then, the first filter is applied and all candidates which overlap the obstacle – or a polygon's hole – are purged. Finally, candidates are filtered so that they are at maximal distance from the border of the polygon or from an obstacle. The last filter stage is not applied yet because the conflict-graph is not completely defined.

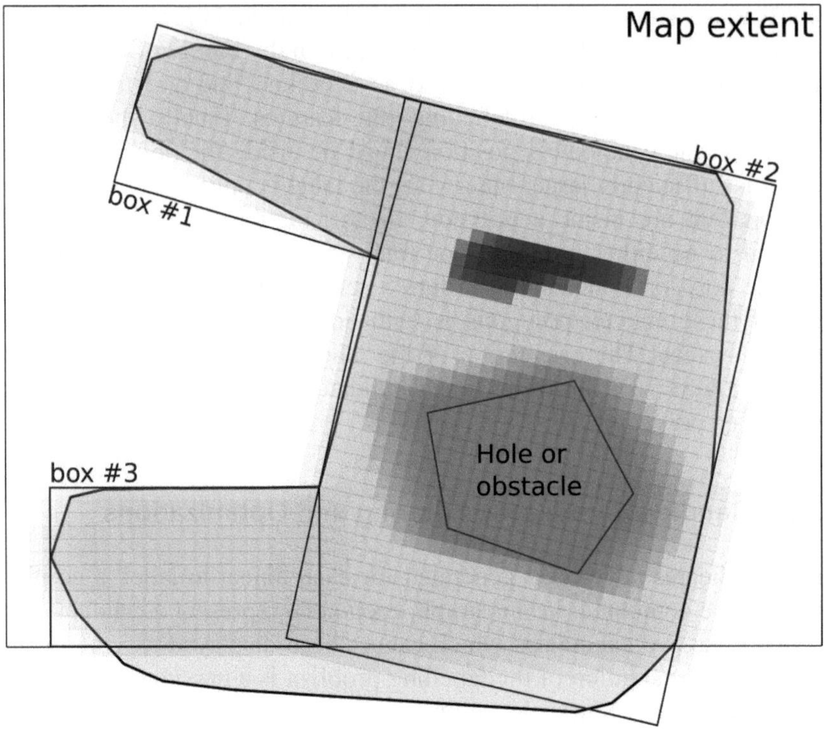

Fig. 10.3. Generation and filtering steps applied to a polygon with an obstacle

Three techniques *chain*, *pop_chain* and *pop_tabu_chain* are available in the library to improve the solution. Moreover, *pop_tabu* (Alvim and Taillard 2008), one of the best techniques available nowadays for point feature label placement is also considered. This technique is based on the generic POPMUSIC optimization frame (Taillard and Voss 2001) and embeds a tabu search method (Glover and Laguna 1997).

10.2.4 PAL Library and Client-Side Integrations

All functionalities are embedded within a C/C++ library. It is released under the conditions of LGPL, GNU Lesser General Public License. Source code and documentation are available through the PAL community portal.[1] Also does the portal provide Trac/SVN instance with bug tracking and a mailing-list to support new integrations and communications between developers willing to improve the library.

---

[1] PAL community portal, viewed January 22nd 2010, http://pal.heig-vd.ch

Using its Java wrapper, PAL has been used to implement extJPAL,[2] an extension for gvSIG 1.x (<1.9). It was a first integration as proof of concept for the project. This extension provides a simple and complete user interface[3] to control all the above described functionalities of PAL and answers the initial purpose to demonstrate its use in real situations. Also, an integration within QGIS in the context of Google Summer of Code 2009[4] has been done, and from version release 1.4 it has PAL inside. Notice that other communities are interesting, like MapWindow, GRASS and MapServer.

10.2.5 Examples

Following examples (Figs. 10.4 and 10.5) have been generated by gvSIG 1.1.2 with the help of the extJPAL extension. The example in Fig. 10.6 has been generated without PAL and shows some bad quality results usually encountered without an intelligent labelling algorithm, like overlapping and label to feature association.

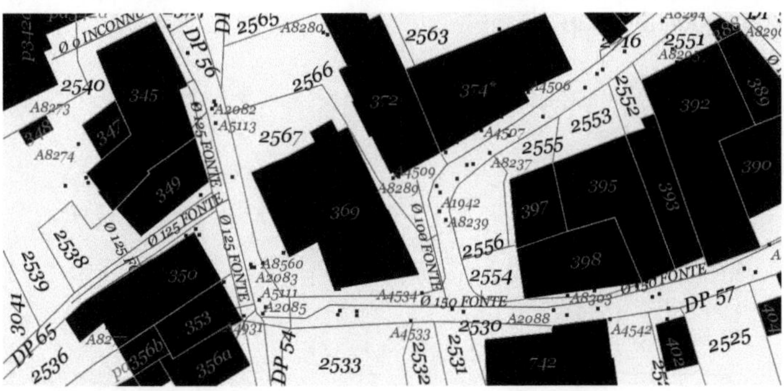

Fig. 10.4. Buildings, parcels water pipes and sewers

---

[2] gvSIG extensions repository, Valencia, Spain, viewed 15 August 2008, https://gvsig.org/plugins/downloads

[3] Screencast of the extJPAL experimental extension for gvSIG 1.1, viewed January 22nd 2010, http://pal.heig-vd.ch/uploads/extJPAL-experimental-screencast.ogv

[4] http://code.google.com/intl/fr/soc/. Accessed 23 Nov 2011

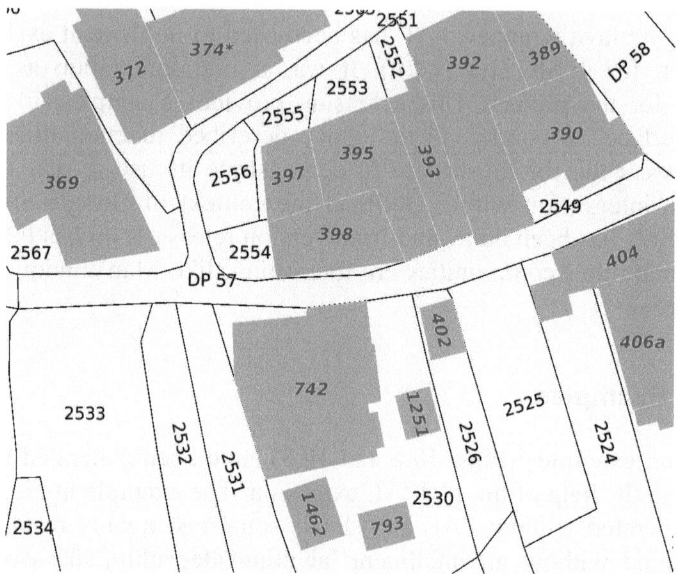

Fig. 10.5. With the PAL library, labels never overlap, and, wherever possible, labels are not displayed over another features than their corresponding one, nor displayed outside the map

Fig. 10.6. Without advanced labelling capabilities, some labels overlap, some describe the wrong feature and some are outside the map

10.3 PAL as a Service

As the PAL library native source code is written in C/C++, any integration for the C world is straightforward. For other languages, a wrapper interface is needed, like JPAL does for Java with Java Native Interface technology, and other wrapper could be written with for instance the help of Python C API, SWIG (Simplified Wrapper and Interface Generator), etc.

The new approach developed within this chapter is to wrap up PAL through a web interface. The availability of a standard-compliant labelling service would help to favour the access to advanced labelling functionalities for any application that need to produce cartographic content of high quality. The next section describes a possible method to realize this wish.

10.3.1 OGC Compliant Labelling Service

The purpose is to expose methods from PAL library through a web service. Let's remind what are the inputs and outputs of PAL library:

- Inputs define the map context and all control parameters:
    - Geospatial features of each layer to label and/or to consider as obstacles
    - Map extent, scale of map, rendering resolution
    - Bounding box for each label
    - Priority for each layer
- Outputs define the global solution:
    - Placement and rotation for each label actually to render
    - Statistic results of the process

These inputs and outputs can be described and formatted in compliance with OGC standards, so as to favour interoperability of systems. This means, any client compliant with standards can take benefit of the service with a minimum of effort. There are four main standards to be considered for a labelling service:

- Geographic Markup Language (GML) (Cox et al. 2004)
- Web Processing Service (WPS) (OGC 2008)
- Symbology Encoding (SE) (OGC 2006)
- Styled Layer Descriptor (SLD) (OGC 2007)

One major difference between a desktop approach and a distributed approach is the on-demand stateless web service call. Next section describes how these standards are used.

10.3.2 Access to Geospatial Features

Desktop context (like for extJPAL) implies a direct access to memory to share features between the GIS client and the PAL library. Distributed context needs a remote web access, which means features have to be transferred between the client and the labelling service.

With the focus of interoperability with standards, the use of GML is then required to encode features. It is a XML grammar to express geographical features. It serves as a modelling language for geographic systems as well as an open interchange format for geographic transactions on the Internet. Nevertheless, depending on the situation, a format like GeoJSON[5] could also be used as it is a wide spread encoding format, especially in geospatial open source communities.

There are two possible sources for geospatial data:

- Local: features from a local layer, stored on the client that calls the service. In this case, the client has to encode features in GML – or GeoJSON – and include them in the service call.
- Remote: features are part of a layer stored on a remote server which is "OGC Web Features Service (WFS)" compliant. In this case, the user will provide a WFS request to the labelling services so as it knows how to get access to these remote features.

10.3.3 Control Parameters

Many parameters are required by the labelling service when a request is sent. Some are explicit and must be provided: property name of the field containing the text, priorities, which layer to consider as obstacles, etc.

Some parameters are implicit and can be inferred to complete the invocation request:

- Map extent can be calculated from the set of features received by the service as only concerned features are received (on-demand call without persistence).
- Bounding-boxes of labels can be computed by the service from label text and their corresponding style – i.e. font, colour, etc. – The text is often stored into a feature attribute, but it can also be created from a template – e.g. for an address: "`{Road name}, {Number}`".

[5] GeoJSON, JSON Geometry and Feature Description, viewed January 22nd 2010, http://geojson.org

To describe all these parameters with an OGC conformance, the use of WPS is a way to go. It is designed to standardize the way GIS calculations are made available over the Internet. It describes any process including all of its inputs and outputs, and triggers its execution as a Web Service. It is first designed to work with geospatial data.

10.3.4 Describing the Labelling Solution

As a "pre-processing service for labelling symbology", output results have to be described. It concerns calculated placements and rotations of labels with their styling properties. The corresponding OGC specifications able to fulfil this need are Symbology Encoding combined with Styled Layer Descriptor. They control the visual portrayal of the geospatial data by the definition of styling rules, so as the map server knows how to render the map during the portrayal process. There are two parts to describe, placements with possibly rotations, and styling properties, which are both described in the followings sections.

10.3.4.1 Definition of an InlineFeature Element

Results are first expressed by label placements, each possibly combined with a rotation. These are a collection of geographical coordinates which represent anchors where to hang up labels. These are kind of "abstract points" and can be encoded with GML points within the *InlineFeature* element described by SLD specification. So, first step, for each label, there is a corresponding *featureMember* with the needed specific details used by the renderer to draw the label: this is the text, its placement and the rotation to apply.

Notice a *Layer* property is also defined to express to which layer the label belongs to. Indeed, the collection contains all labels of the solution. This means if there is more than one layer labeled, labels from all layers are put together in this collection. This property is used by the second step.

10.3.4.2 Definition of the Associated UserStyle

Second step is to describe the symbolization to apply to these inline features. As all features are together in the same collection, the renderer has to filter them to apply the specific symbolizer for each layer. So, there is one *Rule* per layer, and the *Layer* property is used to drive the filtering.

As said, these features are abstract anchors, so only a *TextSymbolizer* has to be applied within each *Rule*. It's the role of the *TextSymbolizer* to draw labels according to the text (*LabelName* property) and the point property placement specified. Notice that Symbology Encoding defines the lower-left corner of the label to hang it up to the placement, and this is exactly what PAL considers while calculating placements. No transformation is needed.

According to type of arrangement (horizontal, free …) there can also be a rotation associated for each label, and thus does the symbology have to process the *Rotation* property if it exists. This is done using the *Rotation* element defined by SE within a *TextSymbolizer*. It defines a clockwise direction rotation in decimal degrees, encoded as a floating-point number. Negative values mean counter-clockwise rotation. The default value is 0.0 (no rotation). Notice that PAL gives the rotation in counter-clockwise radian, so does the service need to do a transformation to conform to SE.

To complete the symbolization description, *Font* and *Fill* styling elements have also to be set for each *TextSymbolizer* (for each layer).

Below. is an example of output for two layers to label:

```
<StyledLayerDescriptor xmlns="http://www.opengis.net/sld"
     xmlns:gml="http://www.opengis.net/gml"
     xmlns:wfs="http://www.opengis.net/wfs"
     xmlns:pal="http://geosysin.iict.ch/pal">
    <Name>Pal Labelling Solution</Name>
    <UserLayer>
        <Name>Anchor Points, rotation and text</Name>
        <InlineFeature>
            <wfs:FeatureCollection>
                <gml:featureMember>
                    <pal:Label>
                        <pal:LabelText>Paris</pal:LabelText>
                        <pal:LabelFID>1</pal:LabelFID>
                        <pal:Rotation>12</pal:Rotation>
                        <pal:Layer>cities:capitals</pal:Layer>
                        <pal:AnchorPoint>
                            <gml:Point>
                                <gml:coord>
                                    <gml:X>2.3333</gml:X>
                                    <gml:Y>48.8667</gml:Y>
                                </gml:coord>
                            </gml:Point>
                        </pal:AnchorPoint>
                    </pal:Label>
                </gml:featureMember>
                <gml:featureMember>
                    <pal:Label>
                        <pal:LabelName>Upernavik</pal:LabelName>
                        <pal:Rotation>42.3456</pal:Rotation>
                        <pal:Layer>cities:others</pal:Layer>
                        <pal:AnchorPoint>
                            <gml:Point>
                                <gml:coord>
                                    <gml:X>-56.15</gml:X>
                                    <gml:Y>72.8</gml:Y>
                                </gml:coord>
                            </gml:Point>
                        </pal:AnchorPoint>
                    </pal:Label>
                </gml:featureMember>
```

```
            <!-- <gml:featureMember>
                     one feature per label of the global solution
                 </gml:featureMember> -->

             </wfs:FeatureCollection>
         </InlineFeature>
         <UserStyle>
             <FeatureTypeStyle
                    xmlns="http://www.opengis.net/se"
                    xmlns:ogc="http://www.opengis.net/ogc">
                 <Rule>
                     <Filter>
                         <ogc:PropertyIsEqualTo>
                             <ogc:PropertyName>
                                 LayerName
                             </ogc:PropertyName>
                             <ogc:Literal>
                                 cities:capitals
                             </ogc:Literal>
                         </ogc:PropertyNameIsEqualTo>
                     </Filter>

                     <TextSymbolizer>
                         <Label>
                             <ogc:PropertyName>
                                 LabelName
                             </ogc:PropertyName>
                         </Label>
                         <LabelPlacement>
                             <PointPlacement>
                                 <Rotation>
                                     <ogc:PropertyName>
                                         Rotation
                                     </ogc:PropertyName>
                                 </Rotation>
                             </PointPlacement>
                         </LabelPlacement>
                         <Font>
                             <SvgParameter name="font-family">
                                 Arial
                             </SvgParameter>
                             < SvgParameter name="font-style">
                                 Bold-Italic
                             </SvgParameter>
                             < SvgParameter name="font-size">
                                 16
                             </SvgParameter>
                         </Font>
                         <Fill>
                             < SvgParameter name="fill">
                                 #ffffff
                             </SvgParameter>
                         </Fill>
                     </TextSymbolizer>
                 </Rule>

            <!-- <Rule>one rule per layer</Rule> -->

             </FeatureTypeStyle>
         </UserStyle>
     </UserLayer>
</StyledLayerDescriptor>
```

The result is a unique SLD/SE description for several input layers. It can be saved to be used several times as soon as the map context stays the same, so as the labelling process doesn't then have to be done each time the map context is viewed.

10.4 Conclusion

More than a description of algorithms (Laurent et al. 2009), PAL is a ready-to-use library for cartographic label placement following a combinational approach with optimization techniques. To enhance labelling capabilities, such a library can be integrated manifold, client-side – e.g. gvSIG and QGIS use PAL – or server-side like for MapServer, but also as a "pre-processing service for labelling symbology". The latter is the main focus of this chapter. It describes a method to wrap PAL as an OGC compliant web service dedicated to labelling. It explains which specifications are used and how to use them to formalize input parameters and results of the library. These are Geographic Markup Language, Web Processing Service, Styled Layer Descriptor, and Symbology Encoding.

The availability of a standard-compliant labelling service help to favour the access to advanced labelling functionalities for any application that needs to produce cartographic content of high quality. The described method can be applied to wrap any labelling library. At the time of the constructions of local, national and global Spatial Data Infrastructures, such a service is required to help the transformation of geodata to knowledge by the production of maps of quality.

It is planned to give a try for such a labelling service. While writing this chapter, a prototype is in development in the context of a work of diploma with a focus on real production environment.

References

Alvim ACF, Taillard ÉD (2008) POPMUSIC for the point feature label placement. Eur J Oper Res 192(2):396–413

Cox S, Daisy P, Lake R, Portele C, Whiteside A (2004) OpenGIS geogra- phy markup language (GML3), Implementation specification version 3.1.0, OGC Doc. No. 03-105r1

Glover F, Laguna M (1997) Tabu search. Kluwer Academic Publishers, Dordrecht

Laurent M, Taillard ÉD, Ertz O, Grin F, Rappo D, Roh S (2009) From point feature label placement to map labelling. In: Proceedings, metaheuristic international conference (MIC'09), Hamburg

OGC (2006) OpenGIS symbology encoding implementation specification (SE), viewed January 20th 2010, http://www.opengeospatial.org/standards/se. Accessed 23 November 2011

OGC (2007) OpenGIS styled layer descriptor (SLD) profile of the OpenGIS web map service (WMS) encoding standard, viewed January 20th 2010, http://www.opengeospatial.org/standards/sld. Accessed 23 November 2011

OGC (2008) OGC(R) Approves Web Processing Service Standard. Press Release on Fri, 2008-02-2

Taillard ÉD, Voss S (2001) POPMUSIC: partial optimization metaheuristic under special intensification conditions. In: Ribeiro C, Hansen P (eds) Essays and surveys in metaheuristics. Kluwer Academic Publishers, Boston, pp 613–629

Wagner F, Wolff A, Kapoor V, Strijk T (2001) Three rules suffice for good label placement. Algorithmica 30:334–349

Wolff A (2010) The map-labeling bibliography. http://i11www.iti.uni-karlsruhe.de/~awolff/map-labeling/bibliography. Accessed 23 November 2011

Part 4: Geospatial Web

Chapter 11 Building Location-Based Mass Collaboration Systems: Challenges and Opportunities

Arie Croitoru[1] and Ofer Arazy[2]

[1]Department of Earth and Atmospheric Sciences and Centre for Earth Observation Sciences (CEOS), University of Alberta, Alberta, Canada. croitoru@ualberta.ca
[2]School of Business (Management Information Systems), University of Alberta, Alberta, Canada. ofer.arazy@ualberta.ca

Abstract. With the rise of Web 2.0 and social computing, we have been witnessing a revolution in the way spatial information is being accessed and utilized. However, the paradigm that governs the generation of spatial information products such as GIS applications has not changed much. The rise of web-based mass collaboration offers an opportunity to develop a new paradigm for creating and sharing spatial information products. To illustrate this, we developed wikiGIS, a location-based mass collaboration system that is designed to generate spatial knowledge products through self-organization, decentralization, and peer-to-peer mass collaboration. Open source software tools provide a natural fit for implementing such location – based mass collaboration systems. In this paper we review the design and development of wikiGIS, analyse the suitability of existing open source tools, and identify key challenges for tools that support location-based mass collaboration systems.

11.1 Introduction

In recent years we have been witnessing a revolution in the way people access and utilize spatial data, information, and knowledge. The availability of the Global Positioning System (GPS), the development of

E. Bocher and M. Neteler (eds.), *Geospatial Free and Open Source Software in the 21st Century*, Lecture Notes in Geoinformation and Cartography,
DOI 10.1007/978-3-642-10595-1_11,

broadband (wired and wireless) communication networks, the emergence of affordable location-aware mobile devices, the growing popularity of geo-browser (e.g. Google Earth and Microsoft Virtual Earth), and the ability to easily mash-up geographic web services (Alameh 2003) have brought personalized spatial information products to the fingertips of everyday users. Within a few years spatial information has become an on-demand, affordable and easy-to-use consumer product.

However, the paradigm that governs the generation of spatial information products has not changed much. Today, the creation of spatial knowledge still relies heavily on a handful of domain experts who are responsible for collecting, analyzing and disseminating such knowledge in highly controlled and centralized organizational frameworks. This approach is rooted in the perception of Geographic Information Systems (GISs) as a complex software dedicated to advanced or expert users. Spatial knowledge is often statically stored in central knowledge bases and cannot be easily modified. Invaluable local spatial knowledge that citizens have plays little, if any, role in the process of creating spatial information products, and the ability of citizens to initiate, influence and contribute to this process is virtually nonexistent. This trend is likely to intensify as it is estimated that by 2011 individuals will be responsible for producing about 70% of the worldwide digital data volume (Chute et al. 2008).

The rise of web-based mass collaboration, in particular in the context of Web 2.0 and peer knowledge production (O'Reilly 2005), offers a tremendous opportunity to transform the way spatial information products are created. Within the Web 2.0 framework, citizens can go beyond the passive role of spatial information consumers and become spatial data creators (Sui 2008), participate in organizing spatial data into information, and contribute to distilling spatial data into a salient body of coherent knowledge. By leveraging online communities and online mass collaboration technologies we have, for the first time, the ability to empower and engage citizens in data gathering, sense-making, and knowledge sharing. Thus, spatial knowledge creation and dissemination can become a dynamic distributed collaborative process.

Given these recent trends and opportunities, we have been developing the concept of location-based mass collaboration systems: a web-based platform for enabling the collection, creation, and organization of spatial knowledge, through the power of large-scale bottom-up citizen collaboration (Croitoru and Arazy 2007). Our goal is to build a spatially-enabled mass collaboration platform that is open and decentralized, yet provides the necessary aggregation mechanisms for creating high-quality information goods. Such a platform could serve numerous applications, which range from participatory journalism and disaster recovery to urban planning, environmental monitoring, and active mobility.

Open Source Software (OSS) can make significant contributions to the development of such systems. Strong evidence to this can be found in a recent white-paper that surveyed the state of major open source GIS software development initiatives (Ramsey 2007), which identified 33 different utilities that range from development libraries to desktop and web applications. As will be discussed in this paper, while this rapidly increasing pool of software tools can already support to some extent the development of location-based mass collaboration systems, further development is needed in order to unleash the full potential of location-based mass collaboration systems. The overarching motivation for this paper is therefore the current gap between the need and availability of well-established theories, models, and tools for developing web-based systems, which supports the creation of spatial knowledge through peer production and mass collaboration processes.

11.2 From Public Participation to Mass Collaboration

Due to the centralized nature of GIS, Public Participation Geographic Information Systems (PPGISs) have been introduced in the 1990s as a tool for increasing public engagement (Schlossberg and Shuford 2005). However, in a PPGIS a designated central authority still control the input, its form, and the outcomes. In contrast, an alternative approach has emerged in recent years – Naïve Geography – that recognizes the importance of common-sense user-level local geographic knowledge (Egenhofer and Mark 1995). This idea is linked to Neo-Geography, in which the roles of data consumers, communicators, and producers are blurred (Goodchild 2007, 2008). The combination of Naïve and Neo-Geography paradigms offer several significant advantages over the existing GIS and PPGIS paradigms. *First*, Naïve and Neo-Geography allow tapping into the inherent spatial intelligence citizens have about their immediate surrounding (Goodchild 2008). *Second*, they enable individuals to contribute up-to-date data and knowledge, thus making the generated knowledge more current. *Third*, if done on a large scale, they enable to harness the wisdom of the crowds in the process of creating spatial knowledge, thus potentially enhancing its quality.

The combination of user-generated data (e.g. so-called “Volunteered Geographic Information” (Goodchild 2007)) and collaborative sense-making has been made possible through the introduction of mass collaboration systems, e.g. wikis, that allow large distributed self-organized online groups to co-create information and knowledge resources. Simply put, a wiki is an enabling tool – a web-based software that allows all viewers of a page to change the content by editing the page online in a browser (Ebersbach et al. 2006). Mass collaboration systems, and in particular wikis, have already made a significant impact. Perhaps the most prominent example to this impact is Wikipedia (Tapscott and Williams 2006; Benkler 2006). Since its creation in 2001, Wikipedia quickly became a worthy rival to traditional encyclopedias that are created by domain experts (Giles 2005). Wikipedia now includes over 10 million entries and attracts over 680 million visitors per year (wikipedia 2009). Similar examples can be found in Open Source Software (Tapscott and Williams 2006; Crowston et al. 2007).

Research on the design, development and utilization of mass collaboration systems for spatial knowledge generation is only in its infancy. Rinner (Rinner 2001) proposed the concept of an Argumentation map where individuals could contribute information to an on-line discussion forum regarding pre-determined geo-spatial data. Sui (2008) suggested that contributions of data need not be limited to location, and can include any georeferenced data. Similarly, Hudson-Smith et al. (2008) and Bishr and Kuhn (2007) suggest that bottom-up GIS created on the same principles of Wikipedia will support the development of better spatial content.

Currently, there is a rapidly growing number of on-line applications that demonstrate some of the principles described above. For example, in MITs "WikiCityRome" Project (2009), citizens serve as voluntary sensors (Goodchild 2007) and fuse on-line collaboration with mobile devices to allow citizens to broadcast their location in real time. In other initiatives, such as Wikimapia (2009) and OpenStreetMap (2009) citizens participate in a collaborative mapping effort in which each mapped location is associated with a wiki page that summarizes the available location-related information. In these applications emphasis is put on collaborative spatial data collection while little attention is given to the process of aggregating low-level data into high-level knowledge through mass-collaboration processes.

11.3 Building Location-Based Mass Collaboration Systems

11.3.1 Design

Location-based mass collaboration systems aim to provide an open decentralized web platform through which aggregation mechanisms for creating high-quality spatial information products are made available. This is accomplished through the fusion of four key elements (further discussion of these elements can be found in Croitoru and Arazy 2007):

- ***Spatial:*** the ability to represent, store, retrieve and query spatial entities (e.g. points, lines, and polygons) and their attributes.
- ***Temporal:*** the ability to associate data entities with a time stamp. Time is essential not only for understanding when things happen, but also for evaluating data relevancy and reviewing revision history. For example, by determining how relevant user contributed data is through a time decay function.
- ***Social:*** the ability to harness mass collaboration depends on a constructive and active user community. Prior mass-authoring projects (e.g. Slashdot, Wikipedia) show that leveraging on the collective wisdom of the masses (Surowiecki 2005), and providing mechanisms for handling conflict and vandalism are essential.
- ***Information aggregation:*** the ability to collect data, organize it into information and synthesize it into salient knowledge. This is accomplished in two stages: (a) Data gathering: allowing a multitude of opinions and views, and encouraging conflict and contradiction. The underlying assumption is that diversity of opinions will result in higher-quality information; (b) Synthesis and sense-making: providing mechanisms that allow users to filter low quality information, synthesize multiple evidences into coherent information, and resolve conflicts.

Based on these four basic elements, location-based mass collaboration systems provide a framework for gathering, combining and synthesizing data from various sources into spatial knowledge (Fig. 11.1). In this framework data from in-situ geo-sensors (Tao et al. 2004) and existing spatial databases can be combined with human observations that are gathered and contributed by participating citizens. Unlike other existing frameworks, location-based mass collaboration systems aim not only to gather data but also to synthesize it into salient knowledge through community-driven processes. Location-based mass collaboration systems therefore enable users to go beyond data collection and play a pivotal role in spatial knowledge creation.

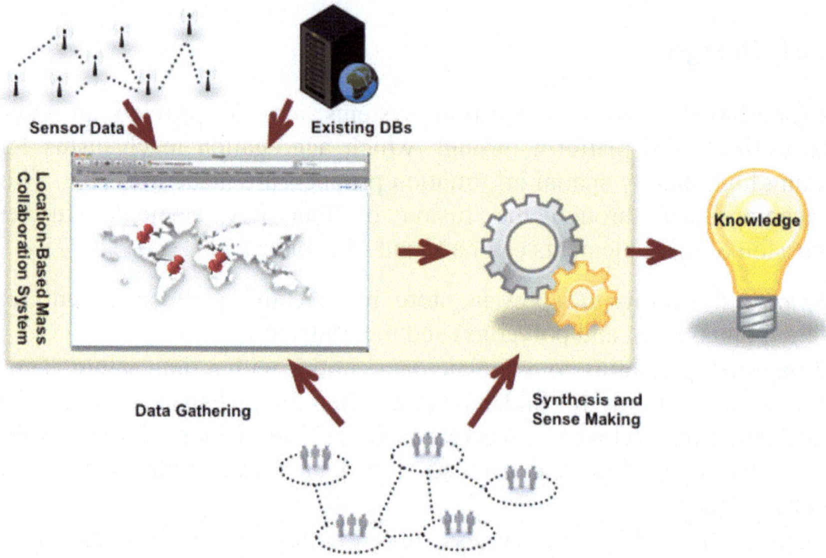

Fig. 11.1. The location-based mass collaboration framework

11.3.2 The Building Blocks

Location-based mass collaboration systems are driven by a bottom-up process of data gathering that is later synthesized to knowledge. Based on this, we distinguish between two constructs, namely *events* and *cases*. An event represents a spatially and temporally tagged observation that is reported by a single contributor. It reflects the experience, opinion, or the data that was collected. From a spatial perspective, an event can be represented by a single geo-referenced spatial entity (e.g. a point, polyline or polygon). From a temporal perspective, an event is associated with a relevancy that decays over time. A case on the other hand, is formed by the aggregation of corroborating events that share a common theme. Cases therefore represent a higher level of aggregation and synthesis of events into a salient knowledge entity. When a case is created, a polygon delineating the extents of the aggregated events is created. Here a polygon represents a schematic outline of the geographic region of the phenomenon described in the case.

To accommodate these different elements location-based mass collaboration systems should wrap together three primary building blocks. First, in order to support the collection of events an environment in which data can be uploaded and presented to the community should be provided. The technology most appropriate for gathering multiple views is a discussion forum, where each event is represented by a thread. This allows discussing the event among community members. By incorporating a rating mechanism, where users rate others postings, we can get an aggregated relevancy score for each event. Rating mechanism can provide the first level of sense making and could help detect hotspots. Second, in order to support the synthesis of multiple events we create a higher-level information object, i.e. a case. This calls for using a technology that supports deliberation. We utilize wiki technology for aggregating citizens' evidence into higher-level synopsis (i.e. a case), and associate a wiki page with each case. Cases inherit and aggregate the relevancy of the events associated with them. Wikis are therefore particularly appropriate since they allow developing mechanisms for deliberation (e.g. discussion pages), conflict resolution and consensus building (e.g. mediation committees, voting), quality control (e.g. watch-lists), and role assignment (e.g. moderators and meta-moderators). Similarly to the existing mechanisms for text revision control and comparison in wiki systems, a case wiki should support spatial revision control, i.e. the ability to go back through the revision history of spatial features and the ability to compare two versions. Third, in order to provide spatial data management, handling, and display a web GIS server should be integrated. The primary role of the GIS server is to manage spatial data storage and retrieval, map rendering, layer management and provide basic viewing capabilities (e.g. pan, zoom or select). Integrating these building blocks into a location-based mass collaboration system can be done through a wrapper. The key role of the wrapper is to coordinate the activity and data flows between the different components, control user authentication, and provide general user interface functionality.

11.3.3 The wikiGIS Platform

Over the last 2 years our group has been developing wikiGIS (Arazy and Croitoru, 2007) as a prototype location-based mass collaboration system for exploring the ideas described above. WikiGIS, which is based entirely on open source software tools, is a web application that wraps together three primary components: a web GIS server (GeoServer 2009), a discussion board (phpBB 2009) and a wiki (Twiki 2009). The web GIS server provides the spatial frame of reference and basic GIS functionality, such as supporting multiple thematic map layers, map display controls (zoom, pan etc.), symbology, and map annotation. In addition, the GIS server is responsible for hosting and maintaining the spatial database. In order to maintain database compatibility, all wikiGIS components are based on the PostgreSQL database engine. The wikiGIS wrapper integrates all three components together into a single web application. The wikiGIS user interface (Fig. 11.2) supports three display modes: an event view, in which only the map and the discussion board are visible; a case view, in which only the map and the wiki are visible; and hybrid view, in which the map as well as the discussion forum and the wiki are visible.

One of the key elements of the wikiGIS wrapper is its ability to seamlessly navigate between the different components. For instance, by selecting a case polygon on the map it is possible to review the corresponding case wiki page, and through that the discussion board events that contributed to the creation of the case. Similarly, by selecting a case in the wiki component it is possible to review both the case on the GIS map and the corresponding events. A unique feature of wikiGIS is that geographic information is fully integrated with both the wiki and the discussion board. Similarly to text editing in a wiki, the spatial representation (i.e. the delineating polygon of a case) can also be edited when a wiki page is revised through a modified wiki revision tracking mechanism.

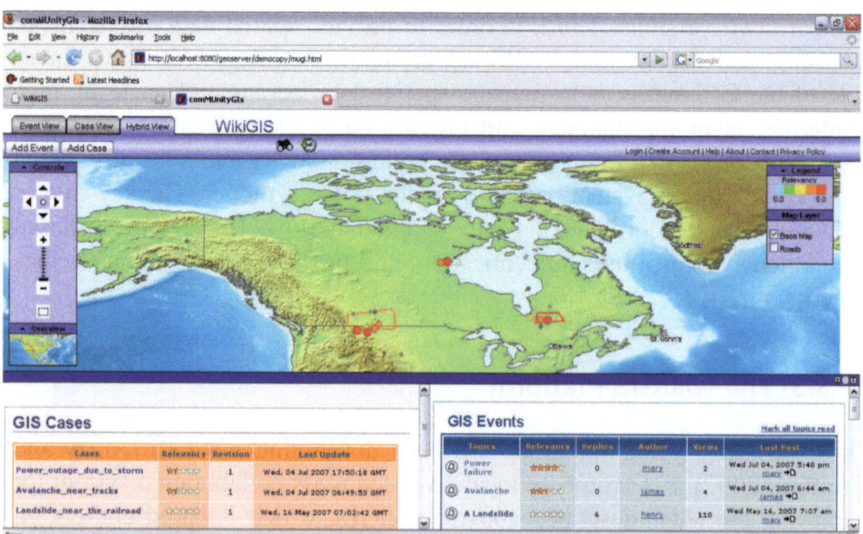

Fig. 11.2. The wikiGIS user interface

11.3.4 Database Integration in wikiGIS

A primary challenge in the design of wikiGIS has been the seamless integration of the different components at the database level. This integration is essential for creating and maintaining a seamless linkage between spatial data, events, and cases, both in space and time. A schematic Entity-Relationship Diagram of wikiGIS is given in Fig. 11.3. WikiGIS is comprised of four different databases that are integrated by the wrapper: the geodatabase database, the wiki database, the discussion board database and the wrapper database. The wikiGIS wrapper database is responsible for managing all the data related to events, cases, and their relevancy rating. The management of events, cases and their association is done in the three relations `events`, `cases`, and `event_to_cases` respectively. Since each event and case is associated with a relevancy measure, the wikiGIS wrapper is also responsible for maintaining event and case relevancy records over time. The wrapper also maintains all user-related information, so that when a user logs into the wikiGIS application she is also automatically logged into the discussion board and the wiki.

The geodatabase (maintained by GeoServer) is responsible to storing all spatial data. When cases and events are created in the wrapper the corresponding spatial entities (i.e. points and polygons) are created in the geodatabase in the `event_features` and `case_features` tables. All spatial data is kept in the ESRI shapefile format. Since wikiGIS is not locked to a specific application domain and is designed to facilitate many different types of spatial events, each event is also associated with a layer that represents a thematic category of spatial features (e.g. oil spills, landslides, etc.). Note that since wikiGIS maintains a full record of the history of spatial features, every revision of a case would create a new spatial feature in the `case_features` table.

Once an event is created in wikiGIS, a new discussion board topic is created in the bbtopics table and the association between the event and the post is stored in the `event_posts` relation. Each post in the discussion board, whether it is an event or a response to an existing event, can be accompanied by additional multimedia files (e.g. images, video clips, voice recordings, etc.) that is stored in `bbposts_attach`. The discussion board entries are kept in the `bbposts` relation and `bbposts_tree`.

The wiki database is similar to the discussion board database. Upon the creation of a case in the wikiGIS wrapper it is linked to a wiki page in the `case_wiki_revision relation`. Users can also add multimedia files (e.g. images, video clips, voice recordings, etc.) to a wiki page. As the case evolves and is revised the revision history is kept in the `wiki_revision` table along with the corresponding time stamp. Note that in the wikiGIS wrapper spatial data is regarded as an integral part of the wiki page data, hence it is possible to edit the case polygon directly from the wiki page.

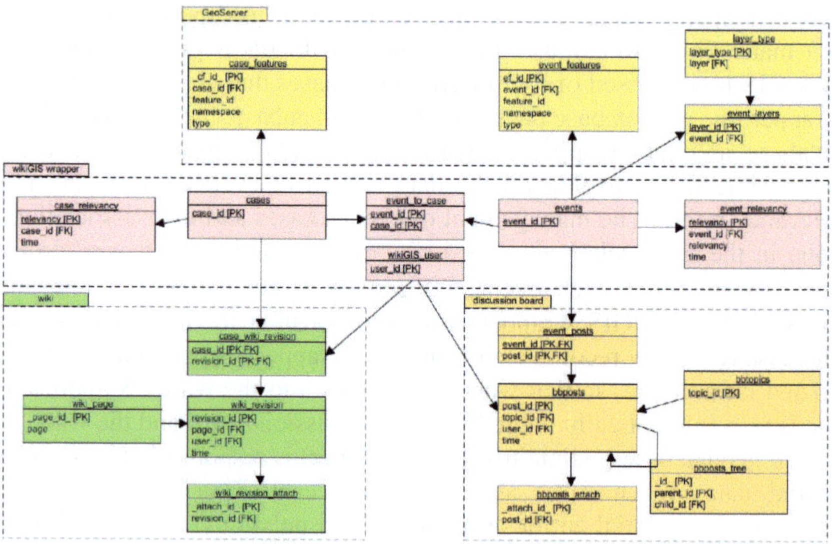

Fig. 11.3. The wikiGIS Entity-Relationship Diagram ([PK] – Primary Key, [FK] – Foreign Key)

11.4 Emerging Challenges

Our experience in designing and developing the wikiGIS platform using open source tools as described in the previous section has highlighted several key challenges that relate to both the theoretical foundations and the technological tools that are currently available for building location-based mass collaboration systems. In this section we discuss these challenges and outline possible future development avenues. It should be noted that the discussion presented here is from a system perspective, i.e. it is focused on developing open source design methodologies and implementation tools. We leave the discussion of the social, behavioral and organizational dimensions for future work.

- *Design patterns*: while there are many grassroots initiatives that aim to leverage web GIS and web 2.0 technologies (e.g. using the Google Maps API or Microsoft Virtual Earth Map Controls SDK), the are often based on an ad-hoc design and there is currently no clear design methodologies for building such applications. Moreover, current mass-collaboration tools (i.e. wiki and discussion board platforms) are often not well integrated with the spatial dimension. There is therefore a need to develop a theoretical foundation for designing and building location-based mass collaboration systems. A first step towards that can be the critical study, systematic mapping and classification of existing initiatives as many of them share some common structure and functionality elements that can be leveraged towards the creation of more general design patterns for location-based mass collaboration systems. A first attempt to conceptualize this idea can be found in (Mader, 2007), which reviews some emerging techniques and constructs that would allow practitioners and technology implementers to design successful wiki-based systems. However, this review is not focused on the spatial dimension.

- ***Collaborative data layers***: in traditional desktop and web GIS systems the organization of spatial data is thematically driven (e.g. roads layer, buildings layer, etc.) and it is typically received from a single source. In the location-based mass collaboration framework data layers are still thematic, but the layer creation process is entirely different: users weave layers together in a distributed bottom-up process by contributing data "patches". The social and collaborative dimensions of the data collection process become important since they play a vital role in the self-organization of the user community, wiki revision control, and have a key role in evaluating data quality. Consequently, a new data model that supports these dimensions should be developed.
- ***Fusing quantitative and qualitative***: the creation of high-level spatial knowledge through large-scale citizen collaboration is based on the synthesis of data from citizens and from existing spatial databases. Some data may be quantitative (e.g. human observations) while other data are quantitative (e.g. an existing map layer or data from in-situ geosensor network; Tao et al. 2004). The first step in generating knowledge from such data is the fusion of the different data sources. One particularly attractive technique for achieving this is the Dempster-Shafer theory (Shafer 1976) due to its ability to incorporate the notion of evidence, and therefore provide a well-structured mathematical framework for incorporated citizen-generate data and existing map data.
- ***Quality mechanisms***: the difficulty of dealing with spatial data quality in location-based mass collaboration systems originates not only from the multidimensionality of the data. The introduction of large-scale user-generated data collection, much of which will be in the form of

qualitative observations, creates significant new challenges for the existing data quality estimation methods and practices (Goodchild 2008). Unlike spatial data collection processes that are carried out by experts who implement strict standards, data contributed by citizens has little or no quality control, and as a result it is difficult to evaluate whether it is trustworthy. Recent work has suggested utilizing trust as a key construct for data quality (Bishr and Kuhn 2007). In light of this, two mechanisms are of particular interest here:

- *Spatio-temporal rating systems*: in typical rating system individual items can be ranked according to how users judge their quality (Segaran 2007). This approach can be further developed to handle spatial data by associating data segments (e.g. segments of a polyline or a region of a polygon) with a quality rating. The temporal domain can also be incorporated by introducing relevance, i.e. applying a decay function to the rating that decreases the elements score as time passes. Relevance was already successfully implemented in a simplistic form in the wikiGIS prototype, and we hope to further develop it in the future.
- *Spatial recommender systems*: the guiding principle behind collaborative recommender (or filtering) systems is the ability to match users (i.e. estimate a similarity score) based on their profile similarity (Segaran 2007). Collaborative filtering can be adopted as an implicit spatial (and spatio-temporal) quality estimation mechanism by expanding the similarity scores currently used to the spatio-temporal domain, e.g. incorporating space and time in the user profile. For instance, given a user interest in a specific area at a specific time interval, the system may recommend to the user data sources that were contributed by other users based on the areas in which these users reside or actively contribute data to.

- ***Text and multimedia georeferencing***: much of the information in the proposed framework is provided in the form of text (i.e. discussion board entries and wiki pages). Although all text in wikiGIS has some relation to a spatial location (e.g. the location of the event), the text may also contain references to many other spatial locations that have a contextual relation to the event or case reported. In order to better exploit the potential of the discussion board and the wiki pages it is necessary to develop tools that would be able to detect such spatial references and their context. This is particularly important in search operations. A possible solution for the development of such geo-referencing capabilities is through the creation of geo-folksonomies (Bishr and Kuhn, 2007), i.e. user-generated gazetteers that would allow users to create and use context-driven georeferencing information.

11.5 Conclusion and Outlook

The ability to organize information and create knowledge products through self-organization, decentralization, and peer-to-peer mass collaboration is already making an impact on our culture, society, and economy (Tapscott and Williams 2006; Benkler 2006). The spatial domain is no exception, and therefore creating a solid theoretical and technological framework for location-based mass collaboration systems is likely to have a significant impact in many application domains that rely on spatial knowledge. Such systems should be seen in the broader context of emerging spatial knowledge networks (John and Melster 2004; Taifi 2008) that would make localized and personalized spatial knowledge available to anyone, anywhere, at any time. Such networks are not centrally controlled – they are in a constant flux through social networks imbedded in spatial domains, and are governed by continuously evolving virtual communities.

This paper reviewed the design and development of wikiGIS, a location-based mass collaboration system prototype. Our work demonstrated how currently existing open source tools (GIS and others) enable creating some of the basic functionality of location-based mass collaboration systems, through a customized wrapper that seamlessly integrates the different system building block. However, we argue that in order to exploit the full potential of location-based mass collaboration systems it is necessary to push the envelope of existing open source tools. In particular, further development of design patterns and support for collaboration-aware data integration and management tools and quality mechanisms are required.

An underlying assumption in this paper has been that the principles of self-organization, decentralization, and peer-to-peer mass collaboration will "naturally" lead to high-quality spatial information products. While there is significant evidence supporting this (e.g. Wikipedia and OSS), research on OSS projects indicates that there is currently only a partial understanding of the nature of these collaboration processes and the settings necessary for their success (Crowston et al. 2007). Moreover, recent work on Wikipedia emphasizes the critical role of governance (i.e. rules and guidelines) in the production of high-quality information products (Viégas et al. 2007). A key challenge in building location-based mass collaboration systems will therefore be the development of a better understanding of location-driven collaboration processes and of adequate spatially-aware governance mechanisms.

References

Alameh N (2003) Chaining geographic information web services. IEEE Internet Comput 7(5):22–29
Arazy O, Croitoru A (2007) wikiGIS: an integrated wrapper for location based mass collaboration. In: Chari K, KumarA (eds) Proceedings of the 17th annual workshop on information technologies and systems (WITS), Montreal, pp 206
Benkler Y (2006) The wealth of networks: how social production transforms markets and freedom. Yale University Press, New Haven
Bishr M, Kuhn W (2007) Geospatial information bottom-up: a matter of trust and semantics. In: Fabrikant SI, Wachowicz M (eds) The European information society – leading the way with Geo-information, Lecture notes in geoinformation and cartography. Springer, Berlin/New York, pp 365–387
Chute C, Manfrediz A, Minton S, Reinsel D, Schlichting W, Toncheva A (2008) The diverse and expanding digital universe – An updated forecast of worldwide information growth through 2011. An IDC White paper
Croitoru A, Arazy O (2007) Introducing location-based mass collaboration systems. In: Chari K, Kumar A (eds) Proceedings of the 17th annual workshop on information technologies and systems (WITS), Montreal, pp 37–42
Crowston K, Li Q, Wei K, Eseryel U, Howison J (2007) Self-organization of teams for free/libre open source software development. Inf Softw Technol 49:564–575
Ebersbach A, Glaser M, Heigl R (2006) Wiki: web collaboration. Springer, New York
Egenhofer MJ, Mark D (1995) Naïve geography. In: Frank A, Kuhn W (eds) COSIT'95, Lecture notes in computer science, vol 988. Springer, Berlin/New York, pp 1–15
Geoserver (2009) URL http://geoserver.org/display/GEOS/Welcome. Accessed 23 November 2011
Giles J (2005) Internet encyclopaedias go head to head. Nature 438:900–901, December 2005
Goodchild MF (2007) Citizens as voluntary sensors: spatial data infrastructure in the world of web 2.0. Int J Spat Data Infrastruct Res 2:24–32
Goodchild MF (2008) Assertion and Authority: The Science of User-Generated Geographic Content. Available at: http://www.geog.ucsb.edu/~good/papers/454.pdf accessed 23 November 2011.
Hudson-Smith A, Batty M, Crooks A, Milton R (2008) Mapping for the masses: accessing web 2.0 through crowdsourcing. Technical report 112, CASA working paper series
John M, Melster R (2004) Knowledge networks – managing collaborative knowledge spaces. In Melnik G, Holz H (eds) Advances in learning software prganization – Proceedings of the 6th international workshop. Lecture notes in computer science, vol 3096, Springer, Berlin, pp 165–171
Mader S (2007) Wikipatterns. Wiley, New York
O'Reilly T (2005) What is web 2.0 -design patterns and business models for the next generation of software, 2005. URL http://www.oreillynet.com/pub/a/oreilly/ tim/news/ 2005/09/30/ what-is- web20.html?. Accessed 23 November 2011
Openstreetmap (2009) URL http://openstreetmap.ca/. Accessed 23 November 2011

Phpbb (2009) URL http://www.phpbb.com/accessed 23 November 2011
Ramsey P (2007) The state of open source GIS. Technical report, Refractions Research Inc.
Rinner C (2001) Argumentation maps: GIS-based discussion support for on-line planing. Environ Plann B Plann Design 28:847–863
Schlossberg AM, Shuford E (2005) Delineating "public" and "participation" in ppgis. urban and regional information systems association. URISA J 16(2):15–26
Segaran T (2007) Programming collective intelligence. O'Reily Media Inc, Beijing/Cambridge/Paris
Shafer G (1976) A mathematical theory of evidence. Princeton University Press, Princeton
Sui D (2008) The wikification of GIS and its consequences: or Angelina Jolie's new tattoo and the future of GIS. Comput Environ Urban Syst 32(1):1–5
Surowiecki J (2005) The wisdom of crowds. Random House, New York
Taifi N (2008) Collaborative knowledge networks. In: Fischer-Hubner S, Duquenoy P, Zuccato A, Martucci L (eds) Proceedings of the Third IFIP WG 9.2, 9.6/11.6, 11.7/FIDIS international summer school on the future of identity in the information society, Springer, Berlin, pp 423–429
Tao V, Liang SH, Croitoru A, Haider ZM, Wang C (2004) Geoswift: open geospatial sensing services for sensor webs. In: Stefanidis A, Nittel S (eds) GeoSensor networks. CRC Press, London, pp 267–274
Tapscott D, Williams A (2006) Wikinomics: how mass collaboration changes everything. Penguin Group Press, New York
Twiki (2009) URL http://twiki.org/accessed 23 November 2011
Viégas FB, Wattenberg M, Mckeon M (2007) The hidden order of wikipedia. In: Schuler D (ed) Second international conference on online communities and social computing, OCSC 2007. Lecture notes in computer science, vol 4564. Springer, pp 445–454
Wiki city rome (2009). URL http://senseable.mit.edu/wikicity/rome accessed 23 November 2011
Wikimapia (2009) URL http://www.wikimapia.org accessed 23 November 2011
Wikipedia, (2009) URL http://en.wikipedia.org/wiki/Wikipedia:About accessed 23 November 2011

[illegible]

Chapter 12 GeoStream: Spatial Information Indexing Within Textual Documents Supported by a Dynamically Parameterized Web Service

Albert Royer[1], Christian Sallaberry[1], Pierre Loustau[1], Mauro Gaio[1], Thierry Joliveau[2], Pierre-André Le Ny[3]

[1]Université de Pau et des Pays de l'Adour, Avenue de l'université, BP 1155, 64013 Pau Cedex, France, {albert.royer, christian.sallaberry, mauro.gaio}@univ-pau.fr
[2]CRENAM-ISIG CNRS/UMR 5600, Université Jean Monnet, Saint-Etienne, 42023 Saint-Etienne Cedex 02, France, thierry.joliveau@univ-st-etienne.fr
[3]Makina Corpus, 30 rue des Jeûneurs, 75002 Paris, France, pierre-andre.le-ny@makinacorpus.com

Abstract. Cultural heritage content is everywhere on the web: digital libraries, archives, and portals of museums or galleries. Cultural heritage document collections are characterized by contents related to a territory and its land's history. In this context, the GeoTopia project is supported by the CNRS-TGE-Adonis and focuses on archive data sharing and interpretation. It consists in a Content Management System (CMS) that aims to manage a repository of multimedia digital documents: it exploits information like origin, theme, period, area, etc. to index and/or query documents. Our contribution is dedicated to spatial information contained in non structured textual documents. More specifically, we have developed a process flow that can extract the spatial information contained in textual documents. This process flow indexes spatial information and computes precise geolocalized representations. We propose to encapsulate it into the GeoStream specific web service and to make its behavior dynamically customizable for easier integration into such platforms used for the management of cultural heritage electronic documents.

E. Bocher and M. Neteler (eds.), *Geospatial Free and Open Source Software in the 21st Century*, Lecture Notes in Geoinformation and Cartography,
DOI 10.1007/978-3-642-10595-1_12,

12.1 Introduction

Managing electronic versions of archive data (histories, travelogues, stamps, maps, etc.) is becoming a task of major importance. In text oriented communities as libraries, archives, museums, etc. the geographic association of information (location, address, place name, etc.) appears to be as important as temporal characterization (Hill 2006). At the same time Geomatics techniques are not any longer limited to topographical entities, material objects or biophysical processes. They are involved in building more general Georeferenced Information Systems including all kind of digital information as Web content, cultural work and even personal life (friends, messages, places, etc.) (Joliveau 2009).

The Geotopia Project (Geotopia is the French acronym for "Geolocalize to transmit, organize, share and interpret archive data") aims to use geospatial techniques for fostering a collaborative enrichment of archive documents. The project is implementing an on-line platform in order to share access, publication and annotation of archive data between archive specialists, scholars and different kind of learned people or technical experts. The project addresses questions related to geoparsing, geovisualisation and collaborative georeferencing. It is also an opportunity to experiment and observe the use of these tools in a real context. One main goal is to understand the role of georeferencing in a collaborative work (Tuffery et al. 2008): metadata management and sharing, gazetteer building, collaborative work organization (by areas, topics, projects, etc.), copyrights management, collaboration between institutions, groups and individuals (amateurs), etc.

The CMS developed in the project supports digitalized archive data publishing, indexing and retrieval. Such data added value relates to local cultural heritage, and therefore, to geographic characteristics. Although well-known search engines still deliver good results of pure key-word searches, it has been observed that precision is decreasing, which in turn means that a user has to spend more time in exploring retrieved documents in order to find those that satisfy the information needed (Kanhabua and Nørvåg 2006). One way of improving precision is to include a geographical dimension into the search as promoted in the GeoTopia project.

During the last 20 years, open source software has undergone an impressive growth in every application domain of Computer Science (Office productivity tools, Databases, Operating Systems, etc.). Among the most noticeable examples, dedicated to NLP (Natural Language Processing), are Linguastream (free of charge licence to academic users, for research purposes), TreeTagger (free of charge licence for evaluation, research and teaching purposes), Gate (licensed under LGPL) prototypes. In the mean time GIS and geo-processing frameworks have been developed (TerraLIB, JGrass, Kalipso, SAGA, SEXTANTE, OrbisGIS, PostGIS, MapServer, OpenLayers, etc.). This evolution gave a new impulse to geomatics. New projects like the GeoTopia one combine NLP tools, GIS functions and geographic resources like gazetteers to support geographic information process flows. In this paper, we are going to focus on a process flow dedicated to automatic geolocalization of spatial information embedded within texts.

Extracting different types of entities from text is usually referred to as Named Entity Recognition (NER). For over a decade, this has been an important NLP task (Chinchor 1998). NER has been successfully automated with near-human performance (Martins et al. 2008). However, the work described here differs from the standard NER tasks for the following reason: the types for our geographic named entities (e.g. references to cities, streets, rivers, mountains) are more accurate than the course-grained types that are generally considered (i.e. person, organization or location). Traditional NER systems combine lexical resources (i.e. gazetteers) with shallow processing, consisting of at least a text segmenter, a lexicon and named entities extraction rules. The extraction rules combine lexicon names with clues like capitalization and surrounding text (Martins et al. 2008). They can be generated by hand or automatically. The former method relies on experts, while the latter uses machine learning with manually annotated training data.

As promoted by (Martins et al. 2008) for non-English languages or very specific tasks, such as the problem of handling thin-grained geographical references, we propose a process flow (Gaio et al. 2008; Sallaberry et al. 2007) based on rules generated by hand.

In order to integrate easily such a geographical information indexing flow in the GeoTopia CMS or in existing library or document management systems, we propose to encapsulate it within the architecture of a web service: GeoStream web service. Moreover, we intend to make its behavior dynamically adaptable. For example, it might be convenient to choose the suitable gazetteer or set of gazetteers when calling the web-service. It might also be interesting to define some degree of priority for such gazetteers invocation according to the characteristics of the text to be indexed.

The paper is organized as follows. In the second section we present related work. In the third section we describe a process flow supporting automatic Geolocalization of spatial information contained within a text. Finally, in the fourth section we describe dynamic adaptation of the behavior of such a geolocalization web service.

12.2 Related Work

Automatic geolocalization of each spatial information embedded within texts relies on geographic named entity recognition. This recognition is the first step to identify the spatial information spread around.

Hereinafter related work points out spatial information is classified into at least two types of spatial information, also called "spatial features" (SF). Simple entities correspond to named entities (e.g. "Paris", "Vignemale peak") and complex entities are derived from simple ones (e.g. "south of Paris", "near Vignemale peak"). We consider simple entities to be absolute spatial features (ASF); whereas complex entities are qualified as relative spatial features (RSF) (Lesbegueries et al. 2006).

Of all the work which focuses on indexing simple entities (Table 12.1), only the GeoSem (Bilhaut et al. 2007) and the PIV (Lesbegueries et al. 2006) (Virtual Itineraries in the Pyrenees) projects deal with complex entities (RSF) during the indexing and the IR (Information Retrieval) phases. The Spirit prototype (Vaid et al. 2005) only handles RSF during the IR phase: it provides a selection list for the user (e.g. "north of", "in", etc.). In fact, all systems, except the PIV, use Minimum Bounding Rectangles (MBR) to represent SF. These produce less precise spatial representations. On the other hand, the PIV prototype uses more accurate representations (points, lines, polygons). All these systems except GeoSem, are supported by the geographical operators and functions proposed by a GIS (e.g. overlapping, intersection, etc.). A variety of spatial IR scores based on overlaps are discussed in (Sallaberry et al. 2008) for these prototypes. Concerning the possibilities of the querying user-interface, it is generally based on keywords. However some systems like GRID (Valcartier 2006) propose a cartographic interface, or like STEWARD (Lieberman et al. 2007) propose a text field for coordinates (latitude/longitude).

Moreover, the STEWARD system (Lieberman et al. 2007) features synthetic views: it uses the frequency of ASF to determine the reference zone associated with each document. The GeoSem project also uses statistical approaches to process geographical queries.

Prototypes / Characteristics	PIV-v1	STEWARD	GRID	SPIRIT	GéoSem
Corpus	Books	Short Text	Web Pages	Web Pages	Long Text
Document returned (document unit)	Paragraphs of a book	//	//	//	Part of a text
Complex entities(RSF)	Index + IR	/	/	IR	Index + IR
Representation	Polygons	MBR	MBR	MBR	MBR
User interface (Kw : keywords, Carto : cartographic interface)	Kw and/or CartoKw and/or Carto	Kw and/or Lat/Long	Carto	Kw and/or Carto	Kw
Statistics	/	X	/	/	x
GIS	X	X	X	X	/

The GeoStream web service we detail in Sects. 12.3 and 12.4 is an extension of the proposal we experimented in the PIV_v1 project.

The OGC (Open Geospatial Consortium[1]) proposes platform-independent standards defining interfaces and operations for data access and manipulation on a set of geographic features. For instance, it proposes Web Feature and Web Map Services dedicated to geographic features management within GIS. Our work is a first step for a similar contribution dedicated to geographic features management within textual digital documents.

Similarly, ViaMichelin,[2] IGN GeoPortail,[3] GoogleMaps,[4] Geonames[5] propose web services that return an address, GPS coordinates or nearby points of interest corresponding to a toponym. Moreover, they may return an itinerary corresponding to two toponyms (departure and arrival). Few works concern services dedicated to text documents. For instance, the Geonames proposes the Wikipedia Fulltext Search web service: it returns the wikipedia entries (as xml documents) found for the searched toponym (place name). The GeoStream web service, described in this article, aims to parse a textual document, to mark, analyze and geolocalize spatial information contained in the document.

The specificities of spatial information evocation within texts, make necessary the use of different parameters and different resources like gazetteers. GeoStream uses reflection to find the right method of the right class to be performed at run time.

---

1 http://www.opengeospatial.org/

2 http://www.viamichelin.fr/

3 http://www.geoportail.fr/

4 http://maps.google.fr/

5 http://www.geonames.org/

Reflection (Sobel and Friedman 1996) is the process by which a computer program can observe and modify its own structure and behavior. It is a particular kind of meta-programming. Among the features provided by a language supporting reflection is the ability to convert a string matching the name of a class or function into a reference to or invocation of that class or function.

12.3 GeoStream: a Web Service Supporting Automatic Tagging, Interpretation and Geolocalization of Spatial Information Contained Within a Text

In this section we describe a process flow supporting automatic tagging, interpretation and geolocalization of spatial information contained within a text. Firstly, we present an example of text. Then, we briefly depict the spatial model on which GeoTopia spatial information process flow relies. Finally, we explain the main stages of the process.

12.3.1 Example

The following example (Fig. 12.1) is an extract of a French text that may be translated as "Jullian reminds of the weapon Factory of Saint-Etienne. It was settled in the Marais quarter in spring 1864." The first stages of our information process flow mark candidate spatial and temporal features. Here (Fig. 12.1), "Julian", "Factory", "Saint-Etienne" and "Marais quarter" are candidate spatial features with respective scores of 0.1; 0; 0.2 and 0.3 whereas "spring 1864" is a candidate temporal feature. These scores (between 0 and 1) evaluate the possibility for each tagged block of text to be spatial (respectively temporal) information.

If we focus on spatial information we can mention that "Marais quarter" is tagged with a stype = 'S' which means that it might be a point of interest. The whole process is detailed in the two next sections.

12.3.2 Spatial Model

The proposed semantic process for analyzing spatial features (SF) relies on an adaptative core model (Gaio et al. 2008). The model is based on a quite naive formal representation of spatial features in comparison with those present in the world of GIS (Clementini et al. 1994; Egenhofer 2002; Hill 2000, GML[6] or WKT[7]).

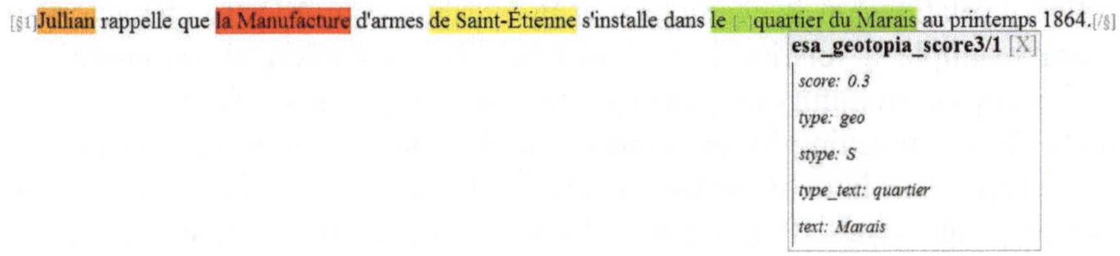

Fig. 12.1. An extract of a French text after making use of a lexical analysis and some spatial and temporal specific grammar rules

It is well appropriate to non-structured textual documents we manage in cultural heritage corpora. According to the linguistic hypothesis, the SF and the temporal features (TF) components may be recursively defined from one or several different SF and/or TF and their relationships; this idea, explained in (Vandeloise 1986), has been easily defined in a recursive way (Fig. 12.2).

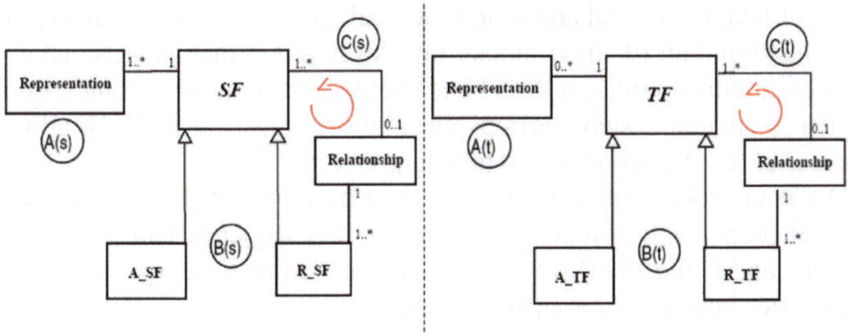

Fig. 12.2. Simplified diagram of spatial and temporal core models

So a SF (Fig. 12.2) has (A(s)) at least a geometric representation. A SF could be (B(s)) an Absolute Spatial Feature (ASF), if it only consists in one named entity allowing a geolocalization. Or a SF could be a Relative Spatial Feature (RSF) if it is defined using a spatial relationship with at least one SF. (C(s)) Spatial relationships can be topological (proximity, inclusion, etc.) or euclidean (distance, geometric, orientation, etc.) (Cohn 1997; Cohn and Hazarika 2001). For instance a proximity relationship appears when we evoke a SF's spatial adjacency to another SF. This relationship is evoked in written language with terms like 'near', 'close by', etc. as "near Saint-

6 Geography Markup Language - http://opengis.net/gml

7 Open Geospatial Consortium - http://www.opengeospatial.org/

Etienne", where the whole expression is a RSF; whereas "Saint-Etienne" is an ASF.

Every relationship is characterized by attributes in order to characterize it. For example a relationship of distance has a numerical parameter; a relationship of proximity (adjacency operator and qualifier) (Sallaberry et al. 2008). All the resulting SF are conform to the spatial core model. Thus, we can manage ASF like "in Saint-Etienne" and relative ones like "near Saint-Etienne", "at about 10 km south from Saint-Etienne", "between Saint-Etienne and Grenoble", etc.

Although this model has been built thanks to linguistic ideas on spatial reasoning, a similar diagram can be modeled for time reasoning (Bessière et al. 1997; Parc-Lacayrelle et al. 2007).

12.3.3 Spatial Information Processing

A document textual content processing sequence is usually composed of four main steps: (1) "tokenization" divides the document into smallest blocks of text, (2) lexical and morphological analysis carries out recognition and transformation of these blocks into lexeme, (3) the syntactic analysis, based on grammars rules, allows bonds between lexeme to be found, finally, (4) the "semantic" step carries out a more specific analysis allowing meaningful lexeme groupings to be interpreted.

Our data processing sequence is a little bit different (Fig. 12.3). After a classical preprocessing textual tokenization (1) sequence and according to (Baccino and Pynte 1994) we adopt an active reading behavior, that is to say sought-after information is a priori known.

A marker of candidate spatial token (2) locates spatial named entities using typographic and lexical rules (involving spatial features initiator lexicons). "stype" characteristics illustrated in Fig. 12.1 are computed during this step. Then, a morphosyntactic analyzer (3) associates a lemma and a nature to each candidate token (i.e. "Marais", noun). A semantic analyzer (4)(6) marks candidate ASF first and candidate RSF next thanks to a Definite Clause Grammar (DCG). For instance, syntagms of composed nouns (i.e. "Marais quarter", "Emile Zola street", "Wild Chamois peak") are brought together (4) and scores are computed (i.e. Fig. 12.1).

ASF are validated and geolocalized (5) thanks to external and/or internal gazetteers (IGN French Geographic Institute resources, Geonames resources and contributive hand-craft local resources) at step five. Then multi-grained expressions containing RSF are built from pointed out ASF (6): embedded spatial relations are interpreted and corresponding geometries are computed (7).

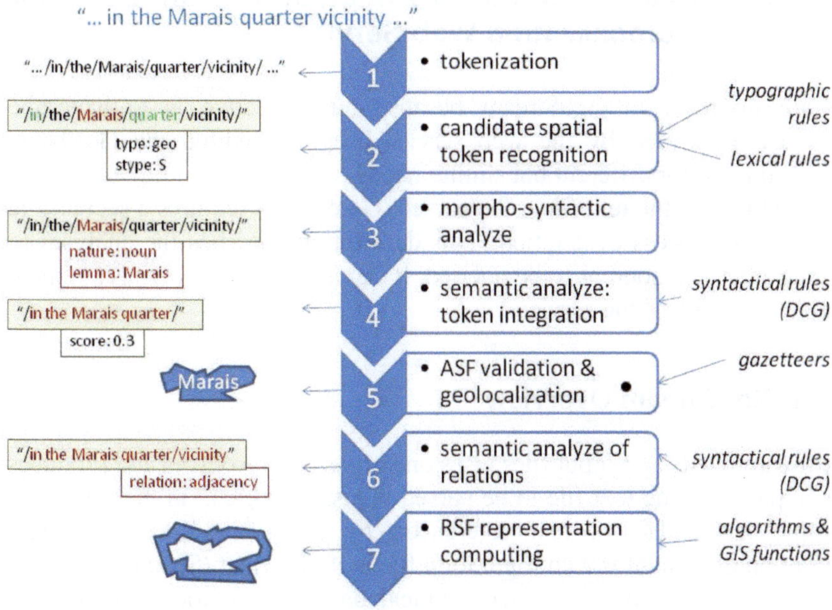

Fig. 12.3. Textual spatial information process flow

This spatial information process flow (Fig. 12.3) is detailed in (Gaio et al. 2008; Loustau 2008). A similar one dedicated to temporal information is detailed in (Parc-Lacayrelle et al. (2007)). Both have been experimented on samples of Pyrenean cultural heritage corpora. GeoTopia project was an opportunity to experiment this approach on new samples of Rhône-Alpes cultural heritage. For an easy integration of the process flow into the GeoTopia platform we encapsulated it within a web service. Currently, the first five stages are integrated into the GeoStream web service so that it tags and geolocalizes the ASF of texts.

Obviously, it might be interesting to indicate which set of rules, which gazetteer (set of gazetteers) is convenient or how many results of geolocalization are required at the most, etc. each time the web service is invocated. This is the reason why we propose a dynamic adaptation of the behavior of such a geolocalization web service. We describe it in the next section.

12.4 GeoStream: a Dynamic Adaptation of the Behavior of a Geolocalization Web Service

We are involved in GeoStream because our objective is to have the processes performed by the web service proposed within the GeoTopia project suitable for different but similar needs.

In addition to the text file to be parsed, GeoStream requires a second file containing a process description. This description specifies the analyses to be done and the resources to be queried. It can be seen, in a way, as a remote configuration file of the service.

12.4.1 GeoStream Overview

GeoStream needs two input files, uses one or several resources and produces one output file. The text file to be parsed is just a text; e.g. in the following example it's a UTF-8 text referenced by an URL. The GeoStream description document is a configuration file in xml format (See Sect. 12.4.3). The resources can be either geographical databases or local contributive gazetteers. The result is an index in xml format (Fig. 12.4).

12.4.2 GeoStream Description Document (GDD)

The GeoStream Description Document states both the process to be performed and the parameters desired. The configuration of the process flow allows for permanent evolution and improvement. The basic point of this configuration is the XML-formatted string sent to GeoStream. The whole process flow is described according to this formalism where the components, their configuration, and the order in which they are invoked, are specified.

In GeoStream, the use of GDD makes it possible to choose (1) the kind of analysis (parsing for Absolute Spatial Features, Relative Spatial Features, Absolute Temporal Features or Relative Temporal Features), (2) which resources are to be consulted and, if any, in which order, (3) what the threshold given as the expected number of results is and (4) which geographic area is concerned.

With the possibilities offered by the java.lang.reflect package, the RunGeoStream (String XMLConfiguration) method, that runs GeoStream, can (1) instantiate the components specified in GDD document, (2) add them to a Stream class instance and (3) execute the flow.

Next, we examine a way to use GeoStream to parse the text showed in Fig. 12.1.

12.4.3 Example of Running GeoStream

Here is the GeoStream description document which specifies (1) the process flow and (2) the parameters to be applied.

```
<?xml version="1.0" encoding="UTF-8"?>
<geotopia-stream>
<composant ordre="1" debug="1"
    class="...geotopia.Composants.Input.InputURL">
    <param nom="Url">http://www.../DGeotopia/jullian.txt</param>
</composant>
<composant ordre="2" debug="1"
    class="...Composants.Linguastream.Linguastream">
    <param nom="Ls_stream_base_dir">.../lib/chaine_ls/</param>
    <param nom="Ls_stream_file">esa.ls</param>
</composant>
<composant ordre="3" debug="1"
    class="...Composants.Georeferenceur.Georeferenceur">
    <param nom="XpathES">//es</param>
    <param nom="XpathESText">./text/text()</param>
    <param nom="XpathESType">./type/text()</param>
    <ressources>
        <ressource
            class="...Composants.Georeferenceur.PostgresGazetteer" ordre="1">
                <param nom="Source">bdtopo_commune</param>
                <param nom="Geotype">P</param>
                <param nom="Host">localhost</param>
                <param nom="Db">geotopia</param>
                <param nom="Query">select distinct nom as NOM, 'P' as TYPE,
                astext(transform(the_geom,4326)) as WKT, canton as CANTON
                from bdtopo_commune where nom ilike '_-_NOM_-_'
                </param>
            </ressource>
            <ressource
                class="...Composants.Georeferenceur.PostgresGazetteer" ordre="1">
                <param nom="Source">bdtopo_lieu_dit_habite</param>
                <param nom="Geotype">S!</param>
                ...
                <param nom="Query">select distinct nom as NOM, 'S' as TYPE,
                astext(transform(geomunion(the_geom),4326)) as WKT,
                nature as NATURE from bdtopo_lieu_dit_habite
```

```
                where nom ilike '%_-_NOM_-_'
                group by nom, nature;
                </param>
30          </ressource>
31          <ressource
                class="...Composants.Georeferenceur.PostgresGazetteer" ordre="1">
32              <param nom="Source">bdtopo_hydronyme</param>
33              <param nom="Geotype">H</param>
                ...
35              <param nom="Query">select distinct nom as NOM, 'H' as TYPE,
                astext(transform(the_geom,4326)) as WKT, nature as NATURE
                from bdtopo_hydronyme where nom ilike '%_-_NOM_-_%
                </param>
36          </ressource>
37          <ressource
                class="...Composants.Georeferenceur.PostgresGazetteer" ordre="1">
38              <param nom="Source">geotopia_rues</param>
39              <param nom="Geotype">R!</param>
                ...
41              <param nom="Query">select distinct nom_actuel as NOM_ACTUEL,
                'R' as TYPE, astext(transform(geomunion(the_geom),4326))as WKT,
                nom_ancien as NOM_ANCIEN, periode_denomination_debut as DE,
                periode_denomination_fin as A
                from geotopia_rues where nom_actuel ilike '%_-_NOM_-_%'
                group by rivoli, nom_actuel, nom_ancien,
                periode_denomination_debut, periode_denomination_fin
                </param>
42          </ressource>
43          <ressource class="...Composants.Georeferenceur.Geonames" ordre="1">
44              <param nom="Parametre">country=FR</param>
45              <param nom="Parametre">style=SHORT</param>
46              <param nom="Seuil">0</param>
47          </ressource>
48      </ressources>
49  </composant>
50  <composant ordre="4" debug="1"
        class="...Composants.XSLTransfo.XSLTransfo">
51      <param nom="XslFilePath">.../lib/xsl/presentationAR2.xsl</param>
52  </composant>
53 </geotopia-stream>
```

In this example, several resources are used. The first category of resources (See lines 16, 23 and 32) is the IGN (French National Geographic Institute) gazetteer. The second one (See line 38) refers to a contributive homemade gazetteer. The third one (See line 43) refers to Geonames. The two first categories are supported by Postgis whereas the last one is supported by Geonames webservices.

This document specifies orders (lines 3, 6, 10, 50), a threshold (line 46) and geographic area (line 44). The components are considered in the order defined by numbers (e.g. ordre= "2"). The number of results required to skip other unexamined resources is given by the threshold (e.g. Seuil > 0). A geographic area can be delimited either by a parameter (e.g. country=France) or by a restriction in the sql query.

With the description document given above, Geostream processes the example (Fig. 12.1). It parses "Jullian reminds of the weapon Factory of Saint-Etienne. It was settled in the Marais quarter in spring 1864. Its first director lived at 26 rue des Mouliniers." and produces the following file:

Date d'interpretation : 2009-12-16
Texte : ***Jullian rappelle que la manufacture d' armes de Saint-Étienne s' installe dans le quartier du Marais au printemps 1864 . Le premier directeur habitait au 26 rue des Mouliniers .***

julliian (de type null et de poids 0.1) : 0 résultat

Saint-Étienne (de type null et de poids 0.2) : 1 résultat

source : bdtopo_commune
geotype : **P**
geoname : **Saint-Étienne**
geometrie : **MULTIPOLYGON(((4.46216741347047 45.3745238187517,4.46194302799851 45.3747323796156,4.46181126553573 45.3748923785765,4.46163025 45.4280063606299,4.24751520804297 45.4284321504819,4.24737039218369 45.4288647718876,4.24736938205707 45.4292427388636)))**
canton : **SAINT-ETIENNE**

Marais (de type S et de poids 0.3) : 2 résultats

source : bdtopo_lieu_dit_habite
geotype : **S**
geoname : **le marais**
geometrie : **MULTIPOINT(4.15418604890164 45.5895089012149,4.17667766648449 45.7401452638789,4.18820657329141 45.7378679047137,4.2203989138224**
nature : **Ecart**
----

source : bdtopo_lieu_dit_habite
geotype : **S**
geoname : **le marais**
geometrie : **MULTIPOINT(4.26536660333024 45.5115074991898,4.32014950090175 45.4409319144101,4.3954906748406 45.4593754964524)**
nature : **Quartier**
----

mouliniers (de type R et de poids 0.3) : 1 résultat

source : geotopia_rues
geotype : **R**
geoname : **Rue des Mouliniers**
geometrie : **MULTILINESTRING((3.89393390644425 27.731721923255,3.89553992390941 27.7322831621504),(3.89242851422763 27.7314001668653,3.892963 27.731499293355,3.89306943629521 27.7315150956988,3.89358699825979 27.7315940969604,3.89364575754826 27.7316172582387),(3.89364575754826 27.7 27.732649653506,3.89566635409963 27.732326836672,3.89557781542964 27.7322958790654,3.89553992390941 27.7322831621504),(3.89553992390941 27.73 27.7323478645917,3.8955842867382 27.7323699002392,3.89554527896313 27.7324282494343,3.8954650409782 27.7325317730791))**
nom_ancien : **Chemin vicinal n°20 des Molières Dufour** *Jusqu'en* **1864**

Fig. 12.4. Geostream result

This result and the corresponding gdd have been obtained after tuning. The initial result and gdd are given in the Annex (Fig. 12.5). Let us comment the final result (Fig. 12.4):

- **Jullian**, former spatial feature candidate, has not been validated; this spatial feature has not been found in any resource.

- **Saint-Étienne**, not typed by the analysis, is validated by the resources; the initial gdd (See Annex) returns three results, the final gdd decreases noise. "ville de Saint-Étienne" which means "city of Saint-Étienne" would have been typed as P (populated place).
- **Marais (quartier du)**, typed by the analysis as S (spot), is validated by the resources; the initial gdd (See Annex) returns nine results, the final gdd partially decreases noise by a restricted search on S geotype.
- **Mouliniers (rue des)**, typed by the analysis as R (road), is now validated by the resources; the initial gdd (See Annex) doesn't return any result because the relevant resource is not invoked (*geotopia_rues* is introduced in the final one).

Note that the Geonames resource has not been used for Saint-Etienne, Marais and Mouliniers parsings because its threshold is fixed to 0 (no more querying for already found features).

12.4.4 Use of java.lang.reflect

As said in the Java documentation, the java.lang.reflect package provides classes and interfaces for obtaining reflective information about classes and objects. The following program uses the class method defined in this package in the same way as the RunGeoStream method.

```
import java.lang.reflect.*;
public class FooTest{
    public static void main (String a [ ])
        throws ClassNotFoundException,
            NoSuchMethodException,
            llegalAccessException,
            InstantiationException,
            InvocationTargetException
            {
        String className = a[0];
        String methodName = a[1];
        Class parameterType [ ]    = new Class [a.length - 2];
        String argument [ ]        = new String [a.length - 2];
        for (int i=2; i<a.length; i++) {
            argument[i-2] = a[i];
            parameterType[i-2] = argument[i-2].getClass();
        }

        // using java.lang.reflect
        Class cl = Class.forName (className);
```

```
21          Method method = cl.getMethod (methodName, parameterType);
22          method.invoke (cl.newInstance (), argument);
23
24          System.out.println("End of FootTest");
25      }
26 }
```

The three lines, which perform the dynamic feature of GeoStream, have to be explained:

- In line 20, the 'forName' method of the predefined 'Class' class returns the class of the argument given by 'className' into the 'cl' variable.
- In line 21, the 'getMethod' method is used to find the method that suits the 'methodName' and 'parameterType' arguments.
- In line 22 the previous appropriate method is sent to an instance of the cl class with the relevant arguments.

Two examples of use are shown below:

```
> java FooTest Foo react1 a_string
Method react1 is running, the parameter is "a_string"
End of FootTest

> java FooTest Foo react2 string1 string2
Method react2 is running, the parameters are "string1" and "string2"
End of FootTest
```

for the given 'Foo' class:

```
public class Foo {
    ...
    public void react1(String S) {
        System.out.println
            ("Method react1 is running, the parameter is \""+S+"\"");
    }
    public void react2(String S1, String S2) {
        System.out.println
            ("Method react2 is running, the parameters are \""+S1+"\" and \""+S2+"\"");
    }
}
```

12.5 Conclusion

The problem of most current CMS integrating the geographic dimension for document indexation and retrieval is that they usually need manual annotations to associate one or more spatial footprints to a document. In this paper we describe an approach dedicated to the automatic indexation and geolocalization of spatial information embedded within texts. We propose and experiment the corresponding GeoStream web service within the GeoTopia CMS. This web service tags, interprets and geolocalizes spatial named entities we call ASF.

We complete this web service behavior with dynamic adaptation possibilities. In such a geographic context, during the indexing stage, it is very interesting to dynamically associate a specific analyzing process flow and specific required resources to each new document stored in the CMS. We use the java.lang.reflect package to experiment and validate reflection through different GDD files specifications.

We extend this spatial information management process flow to the management of complex features (RSF) (Vaid et al. 2005). Future improvement of the GeoStream web service would be to propose and experiment a new version incorporating the management of RSF, of different text formats (UTF-8, ISO-8859-15, etc.).

The final goal is the dynamic choice and running of a subset of steps for the information process flow. As the current choice is limited to ASF, it would be interesting to choose whether to parse ASF, RSF, ATF or RTF (Temporal Features) dynamically in a next GeoStream version.

Acknowledgments

GeoTopia project is funded by the CNRS-TGE-Adonis (NO SUBOS-20.DR7). It is led in partnership with the EVS (http://umr5600.univ-lyon3.fr/) and LIUPPA (http://liuppa.univ-pau.fr) laboratories and the Makina Corpus (http://www.makina-corpus.com/) company.

Glossary

ASF	Absolute spatial feature
ATF	Absolute temporal feature
CMS	Content management system
DCG	Definite clause grammar

GDD	Geostream description document
GIS	Geographic information system
IR	Information retrieval
MBR	Minimum bounding rectangles
NER	Named entity recognition
NLP	Natural language processing
PIV	French acronym for virtual itineraries in the Pyrenees mountains
RSF	Relative spatial feature
RTF	Relative temporal feature
SF	Spatial feature
TF	Temporal feature

References

Baccino T, Pynte J (1994) Spatial coding and discourse models during text reading. Lang Cognitive Proc 9:143–155

Bessière C, Euzenat J, Jeansoulin R, Ligozat G, Schwer S (1997) Raisonnement spatial et temporel Actes 6e journées nationales du PRC-GDR intelligence artificielle

Bilhaut F, Dumoncel F, Enjalbert P, Hernandez N (2007) Indexation sémantique et recherche d'information interactive. In: Proceedings of the CORIA 2007, Saint-Étienne, pp 65–76

Chinchor NA (1998). Overview of Proceedings of the seventh message understanding conference (MUC-7). Fairfax, VA. http://www.itl.nist.gov/iaui/894.02/related_projects/muc/

Clementini E, Sharma J, Egenhofer MJ (1994) Modeling topological spatial relations: strategies for query processing. Comput Graph 18(6):815–822

Cohn AG (1997) Qualitative spatial representation and reasoning techniques. In: KI '97: Proceedings of the 21st annual German conference on artificial intelligence, Springer, London, pp 1–30

Cohn AG, Hazarika SM (2001) Qualitative spatial representation and reasoning: an overview. Fundamenta Informaticae 46(1–2):1–29

Egenhofer MJ (2002) Toward the semantic geospatial web.In: GIS'02: Proceedings of the 10th ACM international symposium on advances in geographic information systems, ACM Press, New York, pp 1–4

Gaio M, Sallaberry C, Etcheverry P, Marquesuzaa C, Lesbegueries J (2008) A global process to access documents' contents from a geographical point of view. J Visual Lang Compu 19: 3–23. Orlando, Academic Press, Inc.

Hill LL (2000) Core elements of digital gazetteers: placenames, categories, and footprints. In: ECDL'00: Proceedings of the 4th European conference on research and advanced technology for digital libraries, Springer, Berlin, pp 280–290

Hill L (2006) Georeferencing, the geographic associations of information. The MIT Press, Boston, p 272

Joliveau T (2009) Connecting real and imaginary places through geospatial technologies: examples from set-jetting and art-oriented tourism. The Cartographic Journal, pp 36–45(10)

Kanhabua N, Nørvåg K (2008) Improving temporal language models for determining time of non-timestamped documents. In: Christensen-Dalsgaard B, Castelli D, Jurik BA, Lippincott J (eds) Proceedings of the ECDL. Lecture notes in computer science, vol 5173. Aarhus, pp 358–370

Lesbegueries J, Gaio M, Loustau P, Sallaberry C (2006) Geographical information access for nonstructured data. In: Haddad H (ed) Proceedings of the SAC, ACM, Dijon pp 83–89

Lieberman MD, Samet H, Sankaranarayanan J, Sperling J (2007) Steward : architecture of a spatio-textual search engine. In: GIS '07: Proceedings of the 15th annual ACM international symposium on Advances in geographic information systems, ACM, New York, pp 1–8

Loustau P (2008) Interprétation automatique d'itinéraires dans des récits de voyages. D'une information géographique du syntagme a une information géographique du discours Thèse de doctorat, soutenue à l'Université de Pau et des Pays de l'Adour

Martins B, Manguinhas H, Borbinha J (2008) Extracting and exploring the geo-temporal semantics of textual resources. In: Proceedings of the IEEE international conference on semantic computing, IEEE, Washington. doi 10.1109/ICSC.2008.86

Parc-Lacayrelle AL, Gaio M, Sallaberry C (2007) La composante temps dans l'information géographique textuelle. Revue Document Numérique 10(2):129–148

Sallaberry C, Baziz M, Lesbegueries J, Gaio M (2007) Towards an ie and ir system dealing with spatial information in digital libraries – evaluation case study. In: Cardoso J, Cordeiro J, Filipe J (eds) Proceedings of the ICEIS 5, Funchal, pp 190–197

Sallaberry C, Gaio M, Palacio D, Lesbegueries J (2008) Extending gis functions for gir needs. In: Proceedings of the ACM 17th CIKM, (GIR workshop), Napa Valley, pp 1–8

Sobel JM, Friedman D P (1996) An introduction to reflection-oriented programming. In: Proceedings of reflection 96, San Francisco, pp 107–126

Tuffery C, Fernandes P, Le Ny P-A (2008) Utilisation d'un site web intégré de webmapping et de gestion de contenus pour la publication de ressources documentaires géoréférencées. Colloque international Le SIG WebMapping dans les sciences archéologiques et historiques, Paris

Vaid S, Jones C B , Joho H, Sanderson M (2005) Spatio-textual indexing for geographical search on the web. In: Medeiros CB, Egenhofer MJ, Bertino E SSTD. Lecture notes in computer science, vol 3633. Springer, pp 218–235

Valcartier (2006) Grid – geospatial retrieval of indexed document. Technical report, R&D pour la défense Canada

Vandeloise C (1986) L'espace en français, Travaux Linguistiques

Annex

Initial gdd

```
<?xml version="1.0" encoding="UTF-8"?>
<geotopia-stream>
    <composant ordre="1" debug="1" class="...Composants.Input.InputURL">
        <param nom="Url">http://www.../DGeotopia/jullian.txt</param>
    </composant>
    <composant ordre="2" debug="1"
                        class="...Composants.Linguastream.Linguastream">
        <param nom="Ls_stream_base_dir">.../lib/chaine_ls/</param>
        <param nom="Ls_stream_file">esa.ls</param>
    </composant>
    <composant ordre="3" debug="1" class="...Georeferenceur.Georeferenceur">
        <param nom="XpathES">//es</param>
        <param nom="XpathESText">./text/text()</param>
        <param nom="XpathESType">./type/text()</param>
        <ressources>
            <ressource class="...Georeferenceur.PostgresGazetteer" ordre="1">
                <param nom="Source">bdtopo_commune</param>
                <param nom="Host">localhost</param>
                ...
                <param nom="Query">select distinct nom as NOM, '?' as TYPE,
                    astext(transform(the_geom,4326)) as WKT from bdtopo_commune
                    where nom ilike '_-_NOM_-_'
                </param>
            </ressource>
            <ressource class="...Georeferenceur.PostgresGazetteer" ordre="1">
                <param nom="Source">bdtopo_lieu_dit_habite</param>
                ...
                <param nom="Query">select distinct nom as NOM, '?' as TYPE,
                    astext(transform(the_geom,4326)) as WKT
                    from bdtopo_lieu_dit_habite where nom ilike '%_-_NOM_-_'
                </param>
            </ressource>
            <ressource class="...Georeferenceur.PostgresGazetteer" ordre="1">
                <param nom="Source">bdtopo_hydronyme</param>
                ...
                <param nom="Query">select distinct nom as NOM, '?' as TYPE,
                    astext(transform(the_geom,4326)) as WKT from bdtopo_hydronyme
                    where nom ilike '%_-_NOM_-_%'
                </param>
            </ressource>
```

```
            <ressource class="...Georeferenceur.Geonames" ordre="1">
                <param nom="Parametre">country=FR</param>
                <param nom="Parametre">style=SHORT</param>
                <param nom="Seuil">0</param>
            </ressource>
        </ressources>
    </composant>
    <composant ordre="4" debug="1" class="...Composants.XSLTransfo.XSLTransfo">
        <param nom="XslFilePath">.../lib/xsl/presentationAR2.xsl</param>
    </composant>
</geotopia-stream>
```

Initial Geostream Result

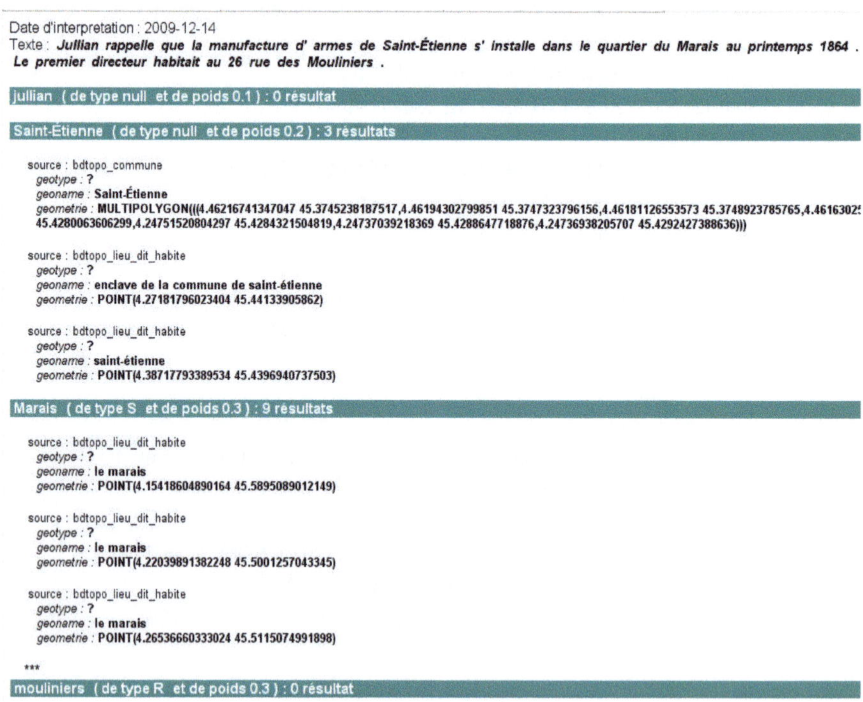

Fig. 12.5. Initial Geostream result

Chapter 13 WIND: an Interaction Lightweight Programming Model for Geographical Web Applications

The Nhan Luong, Patrick Etcheverry, Thierry Nodenot, Christophe Marquesuzaà

IUT de Bayonne Pays Basque, LIUPPA-T2I, 2 Allée du Parc Montaury 64600 Anglet, France, {thenhan.luong, patrick.etcheverry, thierry.nodenot, christophe.marquesuzaa}@iutbayonne.univ-pau.fr

Abstract. Recent research has attested the implementation and the effectiveness of Geographic Information Systems in geographic teaching (Kerski [J Geog 102(3):128–137]; Demirci [Am J Appl Sci 5(3):169–178]). Different works and experiments have shown that current Web Mapping Services and frameworks are partially unfitted for the design and easy programming of web applications dedicated to the teaching of geographic information. Our research problem is enabling teachers to design by themselves an Active Reading Learning Scenario (Murray T (2005) Hyperbook features supporting active reading skills. In: Chapter 8, Web-based intelligent e-learning systems: technologies and applications, pp. 156–174) making use of geographical information avoiding any programmer intervention. In this article, we report on Web Interaction Design (WIND), a web interaction lightweight programming model that we designed to help users to describe interactions between textual, map and calendar components. We present the core concepts of WIND (event-reaction-interaction processing), its API and we illustrate the WIND model with different examples. The main characteristics of WIND are then presented and discussed: WIND is integrative (it combines textual, map and calendar components; it also combines Web Mapping Services, etc.). It is fully executable (thanks to the WIND JavaScript API). WIND promotes lightweight programming. It is object-oriented: users can consistently describe interactions whatever the source and target components are. WIND is a declarative model enabling users to design web interactions

E. Bocher and M. Neteler (eds.), *Geospatial Free and Open Source Software in the 21st Century*, Lecture Notes in Geoinformation and Cartography,
DOI 10.1007/978-3-642-10595-1_13,

between textual, map and calendar components. We present different examples and source codes that pinpoint the added-value of the WIND model and its API.

13.1 Introduction

Today, there is a tremendous amount of web-based cartographic applications that are available to users through proprietary Web Mapping Services (e.g. Google Maps, Microsoft Bing Maps, Yahoo! Maps, France's IGN Geoportail) but also thanks to Free and Open Source Software (FOSS) from the Open Source Geospatial (OSGeo) Foundation: OpenLayers, MapServer, etc. These mapping applications are handled more and more easily thanks to their interaction usage. The main advantage is that the map is updated immediately when users pan around, zoom in or out on a specific area. Moreover, this allows users to search locations, to get directions as well as to add their own information about a place. The term "*interactive map*", therefore, is coined to describe the map with such intuitive interaction techniques put at the users disposal. Most Web Mapping Services provide an Application Programming Interface (API) to support the addition of users-defined overlays (textual or graphical supplements, hyperlinks). However, users require advanced computer science skills to exploit such an API.

More and more educational applications are built around geographic information. Teachers use such applications to teach their learners the concepts of their geographical area, to check understanding of reading texts relating a trip within their region (Nodenot et al. 2006). Some teachers use Web 2.0 tools that have interactions to stimulate the interest of their learners. The communication via such tools also avoids any installation on computers in the classroom.

Our goal is to create applications by describing how the interactions work. The source code of applications should eventually be automatically generated from this description. We are interested in the interaction models allowing to describe the interaction design. Several types of model focus on different aspects of interaction design:

1. Task models describe the activities relevant to the needs of design: GOMS (Goal, Operator, Method, and Selection rules) (Card et al. 1983), UAN (User Action Notation) (Paterno 1999), CTT (ConcurTaskTrees) (Hartson et al. 1990).

2. Dialogue models describe the structure of communication between the users and the application: finite-state machine (Wagner 2006; Petri 1962).
3. Architectural models provide a generic structure of the application from which it is possible to create a particular interactive application: Arch (Bass et al. 1992), MVC (Krasner and Pope 1998), PAC (Nigay 1994).

However, most of interaction models are on high-level abstraction to envisage an automatic code generation.

The research work presented in this paper reconsiders those Web Mapping Services and interaction design models from the viewpoint of educators (e.g. teachers of primary school or secondary school) which try to produce dedicated web applications. They focus on specific localized textual documents called "travel stories" that embed a lot of geographic information about the movements of an actor within a territory. Such geographic information is composed of three complementary features: the spatial feature, the temporal feature and the phenomenon feature (Gaio et al. 2007). In 2007, we developed a first prototype with the Google Maps API coupled with PostGIS (*cf.* http://erozate.iutbayonne.univ-pau.fr/forbes2007/exp/). It is a prototype dedicated to assist a learner for the active reading and better following a travel story on an interactive application. This educational added-value was evaluated in a primary school with two fifth grade classes (9–10 year old children). We set up two specific features of these educational applications compared with currently available web-based cartographic applications:

1. The focus is on interaction between textual, map, and calendar components and not on data visualization.
2. The textual component may display normalized textual documents with the defined CSS styles on tagged words. The map component may display geographic information on a Web-based map with multi-layer mapping services (mainly Google Maps layers, Microsoft Bing Maps layers, and France's IGN layers). The calendar component may display temporal information on an interactive calendar like Google Calendar. The map is no longer the central component, neither is the text, nor the calendar: they play an equivalent role and the user (learner) needs to interact from any of these components and the system should react on any of these components.

Furthermore, this work demonstrated that current Web Mapping Services provide APIs which are partly inadequate to answer 1/ and 2/ of above requirements. This paper will present our results in designing an interaction model and in implementing an API that developers could use to produce web-based mapping services providing end-users with rich interaction abilities.

In the next section, we present the WIND interaction model. In Sect. 13.3, we focus on this API to demonstrate its usability and its advanced potentiality. In the last section, we discuss the advantages and limits of the approach as well as future works.

13.2 The WIND Model

This section presents the WIND model features allowing interactions to be specified and implemented within web applications. We want to highlight that our work focuses on interaction designing rather than interface designing (Beaudouin-Lafon 2004).

We distinguish several types of user who are able to use a web framework to create an interactive application: *developers* who are people that have deep knowledge of Web technology, as well as programming skills; *power users* who do not have significant programming skills but understand the technological difficulties of the problem being treated; and *end-users* who only acknowledge the problem at the design level and do not have any programming skills (Albinola et al. 2009). According to these three types of user, our approach targets currently *developers* and *power users*, but our final goal is *end-users* (teachers and learners).

The web cartographic applications to be implemented should allow users to handle the interactions between its different components such as a map, a text and a calendar. An interaction may be simply defined as a triplet <*area, event, reaction*>. The *reaction* allows the user to perceive the result of his/her action (*event*) on an *area* of the interface. The WIND model has been designed according to this principle.

13.2.1 Example of an Interaction

The application presented on Fig. 13.1 considers a very simple interaction enabling the user to click on the *Anglet* region (in France) of the map (Fig. 13.1a). The system then reacts highlighting all occurrences of the word "*Anglet*" in the text and changing the background colour of the *Anglet* region on the map (Fig. 13.1b).

This interaction can be described in the following way: there is a map area which is sensitive to a user click. This reactive area is decorated with an orange background colour. On a left-click event on this map area, the following changes occur:

- In the text area, the instances of the word "*Anglet*" become decorated with bold, italic, a pink background colour and a blue foreground colour.
- In the map area, the Anglet region is decorated with blue background colour and green border.

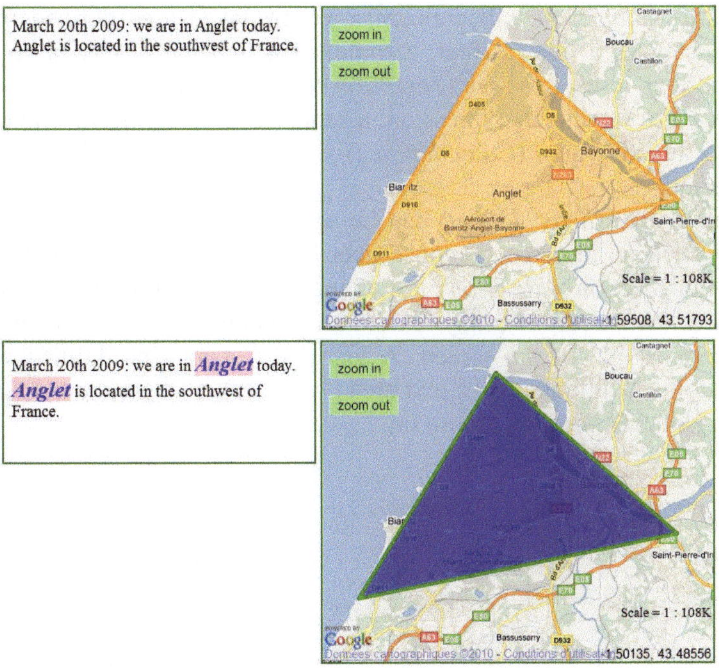

(a) (b)

Fig. 13.1. A simple interaction example

13.2.2 WIND: Core Concepts

WIND (Fig. 13.2) is an operational model that allows both describing and implementing interactions in web applications mixing texts (*Text*), maps (*Map*) and calendars (*Calendar*).

The WIND model relies on the fact that an interactive application presents an interface composed of reactive areas that may trigger system reactions. Thus, an interaction is defined by an area (*SensiblePart*), which, under a specific user action (*event*) will launch system reactions (*Reaction*). A reaction always results (partly) in the modification of an area (*SensiblePart*) of the interface, in order to allow the user to immediately perceive the result of his/her action. In some cases, modified areas can become new reactive area (*SensiblePart*) such as menu opening or button appearance. It is, therefore, able to trigger new system reactions (*Reaction*).

Since WIND describes interactions within applications containing maps, texts and calendars, the reactive areas (*SensiblePart*) on which the user will be able to act could also be text areas (*TextPart*), map areas (*MapPart*) or calendar areas (*CalendarPart*). On a map, for example, a reactive area (*MapPart*) will concern a geographical area defined by a point (a location), a line (a river, a route) or a polygon (a city, a region area). On a text, a reactive area (*TextPart*) could be defined by a particular word or a set of words (an expression, a sentence, a paragraph) while on a calendar, a reactive area (*CalendarPart*) will take the form of a specific date or period.

A reactive area (*SensiblePart*) can be sensitive to various user actions (*event*): a click on the area, a mouse-over or a double-click for example. The system reaction (*Reaction*) following a user action (*event*) on a specific area (*SensiblePart*), may result in a visual effect on one or several areas (*SensiblePart*) of the screen. A visual effect may be defined with different styles: boldfacing for a text area (*TextPart*), background colour of a town on a map (*MapPart*) and so on.

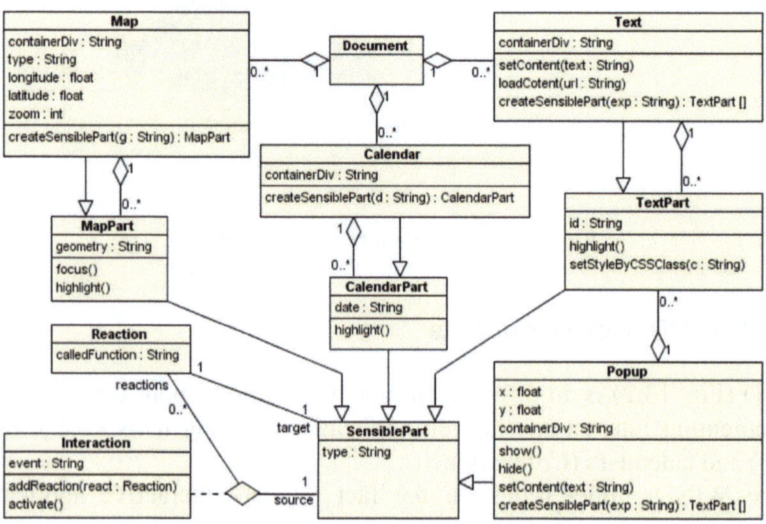

Fig. 13.2. The WIND UML model

13.2.3 Code of a WIND Interaction

The programming languages used for mapping tools play an important role. JavaScript is a programming language for the interaction of HTML and CSS objects in web page; it allows programmers to create new HTML elements, to modify them, to remove them, to associate them with CSS attributes and to deal with events without refreshing the whole web page. Taking advantages of JavaScript, WIND is supported by an API allowing programmers to implement each interaction described at a conceptual level. This API proposes a homogeneous layer built on lower level APIs, specialized in the handling of textual, cartographic and calendar elements. Thus, the WIND functions allow programmers to handle cartographic elements and their associated interactions using open source API such as OpenLayers or proprietary API like IGN Geoportail API. Programmers, however, only handle the functions of the WIND API,[1] they do not need to know the technical features of the lower level APIs called by WIND.

The WIND API integrates the whole model concepts and allows programmers to write their code handling the concepts of interactions, reactive areas, events and system reactions. The WIND API is implemented according to an object-oriented approach allowing programmers to handle each interaction as an object made up of reactive zones, events and system reactions. Thus, coding an interaction deals with building an interaction object composed of a reactive area object (which becomes sensitive to a user event) and one or more system reaction objects.

As an example, the implementation of the interaction described in the Sect. 13.2.1 has been coded following four main steps:

Step 1: Creating the application components.

```
     // Building the text component
var t = new WIND.Text('mytext');
t.setContent("March 20th 2009: we are in Anglet
today. Anglet is located in the southwest of
France.");

     // Building the map component
var m = new WIND.Map('mymap', {'type':
"Google_Street", 'longitude': -1.51, 'latitude':
43.49, 'zoom': 11});
```

Step 2: Creating the reactive areas.

```
     // Building the text sensible parts
```

[1] The WIND API was completely coded from scratch by ourselves. We are currently studying to release it under a BSD license.

```
var tp = t.createSensiblePart('Anglet');
    // Building the map sensible part
var mp = m.createSensiblePart("POLYGON((-1.52
43.53,-1.58 43.46,-1.44 43.48,-1.52 43.53))");
```

Step 3: Creating the system reactions.

```
<style>
.town {
  color : blue; font-weight : bold; font-size :
 15pt;
  font-style : italic; background-color: pink;  }
</style>
var r1 = new WIND.Reaction(mp, 'highlight');
var r2 = new WIND.Reaction(tp,
'setStyleByClass.town');
```

Step 4: Creating the interaction.

```
var i1 = new WIND.Interaction(mp, 'click', null);
i1.addReaction(r1); i1.addReaction(r2);
i1.activate();
```

13.3 Added-Value of the WIND Model

WIND has five main characteristics:

- WIND is integrative.
  By design (*cf.* requirements of web educational interactions taking advantage of route narratives), WIND provides developers with different interaction components. The map is not the central component, neither is the text nor the calendar component. These components play an equivalent role. The user can include within a web application as many instances of each type of component, and we think that WIND could benefit other types of applications in addition to educational ones.
  WIND is also integrative because it offers an overlayer to be compatible with several mapping providers (e.g. Google Maps, Bing Maps, IGN Geoportail) without requiring more programming skills from the users.
- WIND is object-oriented.
  WIND has been designed to take advantages (especially, encapsulation and inheritance) of object-oriented programming language (hence its UML model). The WIND objects are simply created by their class constructor. Methods are disposed to implement relationships between classes. For example, the *createSensiblePart* method of each class (Map, *Text/Popup* or *Calendar*) implements the aggregation relationship

between the *Map* class and the *MapPart* class or the *Text/Popup* class and the *TextPart* class or the *Calendar* class and the *CalendarPart* class; the *addReaction* method implements the association relationship between the *Interaction* class and the *Reaction* class.

The highlight of the WIND model is interaction programming. For this purpose, the *SensiblePart* class plays an important role. It is not only a source reactive area of interaction but also a target reactive area of reaction. Thanks to polymorphism of the *SensiblePart* class (e.g. the *MapPart*, *TextPart*, *CalendarPart* and *Popup* classes inherit from the *SensiblePart* class), the WIND model enables programming interactions for whatever reactive areas. Moreover, interaction programming always has the same structure.

The WIND model allows programmers to add classes inherited from the *SensiblePart* class. These subclasses allow diversifying the reactive areas for interaction. Moreover, the WIND model allows enriching methods of the *Reaction* classes that return visualization as well as improving *event* attribute of the *Interaction* class.

- WIND is executable.

  WIND is not a contemplative design model (Bock 2003; Mellor and Balcer 2002). Thanks to its API, it is an executable web interaction model that enables fast prototyping techniques: developers can rapidly design an interaction and immediately assess it. All instances of the model can be transformed to executable codes. It is very simple to execute because the WIND API has hidden the complex methods. Each reaction is applied to each reactive area thanks to the *calledFunction* attribute when calling the *Reaction* constructor. To simply implement it, the *calledFunction* attribute has a *String* type. For example, if the *calledFunction* attribute is "*hightlight*", then the reactive area will be applied by the *hightlight* method. For this first prototype, we have only supported a few methods for the *SensiblePart* classes.

  The interaction is executable when it is applied by the *activate* method. The reactions defined for the interaction were registered to system and will be executed when an event happens on the reactive area defined for this interaction.

- WIND is declarative.

  Google Maps and OpenLayers programmers will notice that a WIND application has a very simple structure: the JavaScript code of a web interaction includes no conditional or loop statement; there is no event-manager to program. All these elements are devoted to the WIND JavaScript API, and the developer/designer only has to declare sensible parts, events, and reactions.

However, at the present time, using the API to encode interactions still requires some programming skills. Our aim is to enable end-users (particularly teachers) to design such interactions. To reach such an objective, we need to provide them with a dedicated simple authoring environment. The solution that we advocate consists of designing editors that can produce an XML file describing the interaction to be programmed. Currently, we have implemented a first prototype of our toolset that can parse any WIND-compliant XML file in order to generate interactions that the WIND API can execute (*cf.* for example the XML file at http://erozate.iutbayonne.univ-pau.fr/wind/model.xml and its use by an XML parser to produce a web application that the user can exploit: http://erozate.iutbayonne.univ-pau.fr/wind/generator.php?file=model.xml). Such examples show that WIND is declarative.

- WIND promotes lightweight programming.
  The WIND API was programmed with JavaScript language. WIND is executed on the client-side (it is compatible with any up-to-date web browser). This characteristic is quite important because we want teachers to be able to produce WIND applications by themselves and we know that most teachers do not have the skills to install and to maintain server-side frameworks like GeoDjango.[2]

Such WIND capabilities are highlighted by the demonstration available at http://erozate.iutbayonne.univ-pau.fr/wind/. This figure shows the user-interface provided to the end-user: it is an extension of the interaction explained in the Sect. 13.2.1 that exhibits three interacting components: a text, a calendar and a map.

The implementation of these interactions remains simple (*cf.* source code below) owing to the characteristics of WIND. This application was implemented following four main steps (Fig. 13.3):

1. Defining the application components (a text, a calendar and a map)
2. Defining the reactive areas of each component
3. Defining the possible system reactions
4. Building the interactions upon previously defined reactive areas and system reactions

[2] http://geodjango.org/

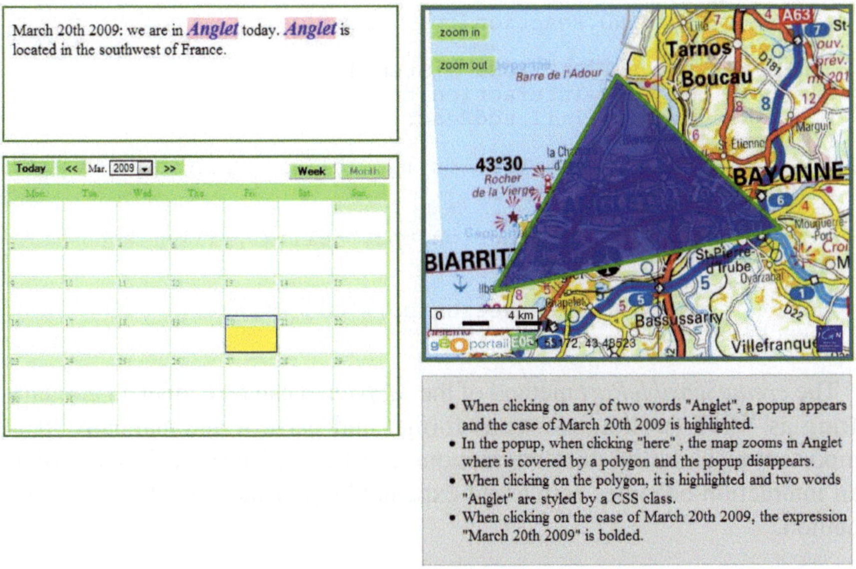

Fig. 13.3. The user interface of a web-based application designed with WIND

```
// DEFINING THE APPLICATION COMPONENTS - Step 1
  // Text component
  var t = new WIND.Text('mytext');
  t.setContent("March 20th 2009: we are in Anglet today.
  Anglet is located in the southwest of France.");
  // Map component
  var m = new WIND.Map('mymap', {'type': "IGN_Street",
  'apiKey': '3*4*********5*4*', 'longitude': -0.24,
  'latitude': 44.4, 'zoom': 6});
  // Calendar component
  var c = new WIND.Calendar('mycalendar');

// DEFINING THE REACTIVE AREAS - Step 2
  // Text reactive areas
  var tp = t.createSensiblePart('Anglet');
  var tp1 = t.createSensiblePart('March 20th 2009');
  var p = new WIND.Popup(650, 240, 'mypopup');
  p.setContent("Click here to zoom in Anglet.");
  var tp2 = p.createSensiblePart('here');
  // Map reactive area
  var mp = m.createSensiblePart("POLYGON((-1.52 43.53,-1.58
  43.46,-1.44 43.48,-1.52 43.53))");
  // Calendar reactive area
  var cp = c.createSensiblePart("2009-03-20");

// DEFINING THE SYSTEM REACTIONS - Step 3
  var r1 = new WIND.Reaction(mp, 'focus');
  var r2 = new WIND.Reaction(mp, 'highlight');
  var r3 = new WIND.Reaction(tp, 'setStyleByClass.town');
  var r4 = new WIND.Reaction(p, 'show');
  var r5 = new WIND.Reaction(p, 'hide');
  var r6 = new WIND.Reaction(cp, 'highlight');
```

```
  var r7 = new WIND.Reaction(tp1, 'bold');

// BUILDING THE INTERACTIONS - Step 4
  var i1 = new WIND.Interaction(tp, 'click', null);
  i1.addReaction(r4); i1.addReaction(r6);
  i1.activate();
  var i2 = new WIND.Interaction(mp, 'click', r2);
  i2.addReaction(r3);
  i2.activate();
  var i3 = new WIND.Interaction(tp2, 'click', r1);
  i3.addReaction(r5);
  i3.activate();
  var i4 = new WIND.Interaction(cp, 'click', r7);
  i4.activate();
```

The *createSensiblePart* method of the *Map* class can take input a geometry string as WKT (well-known text) format and parse it automatically. It is important to underline that interactions can be implemented in any order. An interaction becomes active and executable after the call of the *activate* method.

13.4 Discussion and Future Directions

We presented WIND, a computational model for interaction lightweight programming of geographical web applications. We showed that the WIND model is fully executable thanks to the API that we provide to developers. This API enables developers to describe interactions between Map / Text / Calendar components without having to know anything about the underlying Web Mapping Services (IGN Geoportail API, Google Maps API, OpenLayers API, etc.).

We are currently working on the following improvements. In this current version of WIND, we only consider interactions in which the system reacts immediately to the user's action. A system reaction can be either simple (Fig. 13.4a) or composed of several reactions (Fig. 13.4b).

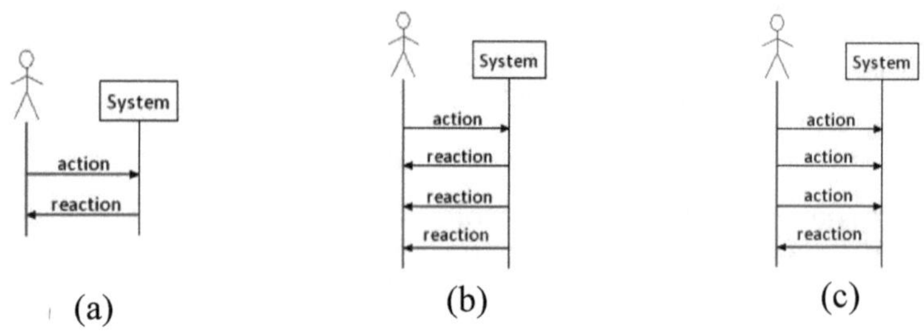

Fig. 13.4. Different kinds of interactions

We have identified three main areas for improvement:

- The WIND model does not allow developers programming interactions where the system reacts after a sequence of several specific user actions (Fig. 13.4c). To take into account this kind of interaction, we plan to improve the WIND model by integrating the concept of an aggregated event. An aggregated event would be triggered when the user raises a set of events (ordered-set, unordered-set, etc.).
- We also plan to extend the reactive area concept because, currently, the WIND model only considers areas of low semantic level. Defining a reactive area on a map only consists in defining a point, a line or a polygon but we plan to extend the reactive area definition by introducing higher level concepts such as city or river for a map area. This work consists of specializing the *SensiblePart* class and improving the WIND API in order to integrate geographical methods allowing to localize higher level areas on a map.
- We also consider that the model could integrate some temporal dimension of both system reactions and user events. In the first case, this means proposing temporal operators allowing the programmer to define the chronology of system reactions resulting from a user event. In the second case, the goal consists in using these temporal operators to specify temporal sequences of user events that will define an aggregate event able to launch one or more system reactions.
- Such improvements would be very useful to describe educational interactions.

Finally, one of our main aims is to enable teachers (i.e. *end-users*) to take into account a particular route narrative, from its first interpretation by the PIIR (*Prototype Interprétation Itinéraires dans des Récits*) toolset (Loustau et al. 2008a ; 2008b and http://erozate.iutbayonne.univ-pau.fr/Nhan/piir/) to the design of a learning scenario, taking advantage of the geographic semantics embedded in such a document. This will require further improvements of our toolset and will lead us to implement a visual programming environment using the WIND API.

Acknowledgments

This research is supported by the French Aquitaine Region (project n°20071104037) and the Pyrénées-Atlantiques Department ("*Pyrénées: Itinéraires Educatifs*" project).

References

Albinola M, Baresi L, Carcano M, Guinea S (2009) Mashlight: a lightweight mashup framework for everyone. In 2nd Workshop on Mashups, Enterprise Mashups and Lightweight Composition on the Web Madrid, Spain April 20th, 2009 Accessed 23 November http://www.integror.net/men2009/papers.php

Bass L, Little R, Pellegrino R, Reed S, Seacord R, Sheppard S, Szczur MR (1992) The UIMS tool developers' workshop: a metamodel for the runtime architecture of an interactive system. SIGCHI Bulletin 24(1):32–37

Beaudouin-Lafon M (2004) Designing interaction, not interfaces. In: Proceedings of international working conference on advanced visual interfaces, AVI 2004. Gallipoli (Italy) ACM, pp 15–22

Bock C (2003) UML without pictures. IEEE Comput Soc Press 20(5):33–35

Card SK, Moran TP, Newell A (1983) The psychology of human-computer interaction. Lawrence Erlbaum Associates, Hillsdale

Demirci A (2008) Evaluating the implementation and effectiveness of GIS-based application in secondary school geography lessons. Am J Appl Sci 5(3):169–178

Gaio M, Sallaberry C, Etcheverry P, Marquesuzaà C, Lesbegueries J (2007) A global process to access documents'contents from a geographical point of view. J Vis Lang Comput 19(1):3–23, ISSN:1045-926X

Hartson HR, Siochi AC, Hix D (1990) The UAN: a user-oriented representation for direct manipulation interface designs. ACM Trans Inf Syst 8(3):181–203

Kerski JJ (2000) The implementation and effectiveness of geographic information systems technology and methods in secondary education. J Geog 102(3):128–137

Krasner GE, Pope ST (1988) A cookbook for using the model-view-controller user interface paradigm in Smalltalk-80. J Object Oriented Program 1(3):26–49

Loustau P, Nodenot T, Gaio M (2008a) Spatial decision support in the pedagogical area: from travel stories to geocoded itineraries. In: Proceedings of 6th international conference on human system learning. Europia IEEE, Toulouse, pp 14–21, 2008

Loustau P, Nodenot T, Gaio M (2008b) Spatial decision support in the pedagogical area: processing travel stories to discover itineraries hidden beneath the surface. In Proceedings of the 11th AGILE international conference on geographic information science (AGILE 2008) The european information society: Taking geoinformation science one step further, pp 359–378, LNG&C, Springer

Mellor S, Balcer M (2002) Executable UML – A foundation for model-driven architecture. Addison-Wesley, Boston

Murray T (2005) Hyperbook features supporting active reading skills. IN: Chapter 8, Web-based intelligent e-learning systems: technologies and applications, pp. 156–174

Nigay L (1994) Conception et modélisation logicielles des systems interactifs: application aux interfaces multimodales. Thèse de doctorat, Université Joseph Fourrier, Grenoble 1

Nodenot T, Loustau P, Gaio M, Sallaberry C, Lopisteguy P (2006) From electronic documents to problem-based learning environments: an ongoing challenge for educational modeling languages. In: 7th international conference on information technology based higher education and training, pp 75–86, 10–13 Sydney, Australia, July 2006

Paterno F (1999) Model based design and evaluation of interactive application. 1st Edition., Series Applied Computing, Springer Verlag

Petri CA (1962) Fundamentals of a theory of asynchronous information flow. In: Proceedings IFIP congress, North-Holland Publishing Company, Amsterdam, 1962

Wagner F (2006) Modeling software with finite state machines: a practical approach. Auerbach Publications, New York

Part 5: From Feedback to Future

Chapter 14 Free and Open Source GIS: Will There Ever Be a Geo-Linux?

Gilberto Câmara, Lúbia Vinhas, Ricardo Cartaxo Modesto de Souza

National Institute for Space Research, INPE, Av dos Astronautas 1758, São José dos Campos, SP, Brazil
gilberto.camara@inpe.br, {cartaxo,lubia}@dpi.inpe.br

Abstract. This paper examines the constraints that limit the large-scale adoption of open source GIS. Although the open source GIS community has already achieved relevant results, their products have a small market share. There is equivalent to Linux and Apache in the open source GIS scene. We try to explain why this happens, by considering some factors that control the evolution and adoption of open source software. Our view is that the community effort is split in many different systems, not allowing a dominant solution to come forth. Thus, none of the current open source GIS has the potential to be a disruptive technology. Then, we consider a future scenario where most public geospatial data will be available as open access policy. This scenario is becoming more probable given recent data policy decision in Europe, USA and other countries. In this scenario, there is a major chance for a disruptive open source GIS to appear.

14.1 Introduction

Free and open source software (FOSS) has moved from being a specialist niche to the mainstream. The FOSS community has grown considerably, as illustrated by its multiple workshops and meetings. In some areas of information technology (supercomputers, data servers, and scripting languages) the leading market solutions are open source. Supercomputers and data servers use Linux as their operating system and companies such as IBM have hundreds of programmers in their Linux teams. All major

E. Bocher and M. Neteler (eds.), *Geospatial Free and Open Source Software in the 21st Century*, Lecture Notes in Geoinformation and Cartography,
DOI 10.1007/978-3-642-10595-1_14, © Springer-Verlag Berlin Heidelberg 2012

scripting languages (Perl, Python, Lua, Ruby) are FOSS. However, there are some areas where FOSS has failed to gain a substantial market share. Document production software is dominated by a single proprietary solution, despite the community efforts for building solutions such as OpenOffice. Nagy et al. (2010) examine the rate of adoption of FOSS in the business community. Their estimates indicate that the Apache web server has about 50% market share and Linux is used as the operating system of about 30% of servers. By contrast, the Open Office suite has a marginal market share of about 3% of the desktop. As the authors state: *"This pattern suggests that there may be significant barriers to open source software adoption among some sectors of the user populations"*. Clearly, some applications areas are more prone for FOSS adoption than others.

In the area of GIS (geographical information systems), the commercial market is dominated by one company (ESRI), which has more than 30% of market share considering software and services. By contrast, the free and open source GIS scene (FOSS4G) is fragmented in diverse niches. Recent surveys of open source GIS provide a good overview of the different products (Steiniger and Bocher 2009; Hall and Leahy 2008). The view expressed in these papers is generally positive of the accomplishments of OSS GIS community. Ramsey (2007) remarks that *"existing products are now entering a phase of rapid refinement and enhancement"*. Steiniger and Bocher (2009) point out that *"[FOSS4G] projects have reached a mature stage and the software offers a multitude of functionalities."* Nevertheless, none of the many available open source desktop GIS has a significant market share. In this chapter, we examine the reasons that limit large-scale adoption of FOSS4G. Thus, the question we address in this paper is: *will there ever be an FOSS4G equivalent to Linux or Apache? If so, how will a Geo-Linux appear?*

Thus, this paper is concerned with examining the structural constraints that limit the adoption of FOSS4G by a large number of users. We start by considering some factors that control the evolution and adoption of FOSS. By examining the reasons why Apache has 50% market share and OpenOffice only 3% (Nagy et al. 2010), we can learn some lessons which are useful for the future of FOSS4G. Then, we try to explain why the FOSS4G scene is split up in many different systems. Finally, we propose some possible futures for FOSS4G development.

14.2 Factors for User Adoption of FOSS

This section proposes some structural reasons that influence users to select a FOSS product. There are many papers that analyze the nature and evolution of FOSS (Nagy et al. 2010; O'Reilly 1999; Ghosh et al. 2002). Most of these papers discuss the organization and motivation of FOSS teams and the economics of open source software development. We believe that the structural reasons that govern the users' adoption of a FOSS have not been discussed in detail in the literature (Câmara and Fonseca 2007). In the following discussion, we focus on the factors that users need to consider for adoption of FOSS. On purpose, we left out an obvious factor: lower cost. Although promoted by FOSS advocates as a major drive for adoption of open source technology, lower cost by itself is not sufficient for users in most cases. This is verifiable empirically by considering the low adoption rate of OpenOffice. Thus, we propose the following factors as contributing to the adoption of FOSS.

Degree of modularity. One of the major factors mentioned in favour of open source adoption is capacity of users to modify a FOSS to suit their needs (Weber 2004). In practice, such adaptability is not easy to achieve since it requires a modular software organization. The more a FOSS is modular, the easier it is to adapt it to the needs of different organizations. Each software product has a periphery to kernel ratio that constrains the potential for modularization, since the kernel needs a tightly organized and skilled programming team. An operating system such as Linux has a well-defined kernel for process control and a periphery consisting of programs such as device drivers, applications, compilers and network tools. Differently, database management systems have a kernel of integrated functions (parser, querier, scheduler, and optimizer) and a much smaller periphery. A high periphery to kernel ratio means a more modular software. The more the kernel is dominant in a software product, the less modular it is, and the less potential contributors it has. One good example is the R suite of statistical tools (Ihaka and Gentleman 1996), whose kernel consists of a small and well-designed library to deal with scientific data. Developing a new application using R is very simple. Many different researchers have developed a large set of statistical software that makes R very popular worldwide.

Shared conceptualization. A good conceptual design is a crucial part of any successful software project. Faulty designs are a major cause of failures in software projects (Brooks 1987). The design problem is even more important for FOSS. If programmers scattered in different places are to communicate well, they need to share the same conceptual view. This conceptual view is difficult to capture in written documents, and is much easier to achieve when there is a prior common background. This explains why many successful projects rely on existing designs. We call this 'shared conceptualization' (Câmara and Fonseca 2007). Again, the R suite of statistical tools is a good example. R has a simple and well-understood conceptual design, based on the S-PLUS proprietary product (Chambers 1998). Their innovative contribution lies not in their design, but on the analysis functions that scientists develop using these environments. The two main conditions for shared conceptualization to happen are:

1. The post-mature perspective (Câmara and Onsrud 2004): a private company develops a software product, for which it holds the intellectual property rights. As the product becomes popular, its functionality and conceptual model becomes well settled, and it becomes part of the "public commons". The popularity and usability of the software motivates other institutions to develop a public domain equivalent, as in the case of R.
2. The standards-led perspective (Câmara and Onsrud 2004): standards consolidate a technology and allow compatible solutions from different producers to compete in the marketplace. An example is the SQL database standard, which has motivated products such as mySQL and PostgreSQL. Another example is the POSIX standard for operating systems, which has served as guidance to Linux.

Design and maintenance challenge. Many users and companies use FOSS because it would be hard for them to design and maintain a proprietary solution. For example, consider a hardware company that needs an operating system and has to choose between building its own or adopting an OSS. Knowing that the Linux kernel has over 12 million lines of code and works well, the company's managers have a large incentive of using Linux. Indeed, the high cost of maintaining a proprietary Unix-like operating system is the main reason why IBM, SGI, and HP switched to Linux. Munga et al. (2009) point out that IBM has made FOSS a core part of their business strategy, and there are an estimated 600 programmers working full-time as part of their Linux team. The same factor explains the adoption of the Apache web server. Designing and maintaining a reliable and fast web server is difficult. The Apache HTTP Server today powers nearly 112 million websites worldwide. Apache's proven reliability and large user base deters commercial companies or other OSS developers from developing alternatives. Thus, when an OSS provides a service that would be hard or costly to replicate, users have a large motivation for using it, instead of developing their own solutions or using proprietary software.

Code stability and avoidance of forking. Programmers use the word "forking" to describe the situation when different groups develop incompatible products, starting from the same code. The avoidance of forking and the resulting code stability is an important factor in FOSS adoption by users. To show the negative impact of forking, consider the parallel evolution of Linux and FreeBSD. From 1977 to 1995, the Computer Systems Research Group (CSRG) of the University of California at Berkeley developed a Unix-like operating system, called the Berkeley Software Distribution (BSD). After the University stopped this activity in 1995, the code was released as open source and taken up by a community of programmers. Differences of opinion among the code developers led to fragmentation of the BSD-Unix community into FreeBSD, OpenBSD, and NetBSD camps (Nagy et al. 2010). By some accounts, the BSD-derived operating systems were more advanced than Linux in the early 2000s (Yu et al. 2006). However, the disputes between the three BSD communities were damaging to their reputation. By contrast, Linus Torvalds has always maintained a tight control over the Linux kernel, an attitude that ensured code stability and user trust. Forking is thus one of the reasons why BSD failed to reach a wide user base, while Linux emerged as the leader of the open source operating systems (Nagy et al. 2010).

Path Dependence and Lock-in Effect. Path dependence is an established property of technology. Although definitions vary, the idea of path dependence is that the current status of a technology (e.g., the dominant software at the desktop) has evolved as a consequence of the initial configuration (e.g., IBM's selection of Microsoft for developing the operating system of its PC) (David 2007). According to Page (2006) path dependence has different causes including *increasing returns*, *self-reinforcement, positive feedbacks and lock-in*. Increasing returns means that the more a choice of a product is made, the greater its benefits both to the producer and consumer. Self-reinforcement means that when a costumer makes a choice, he gets connected to other costumers that encourage that choice to be sustained. With positive feedbacks, the benefit of making a choice is increased when others do the same. The "lock-in" effect occurs when a customer is dependent on a single supplier for products and services and cannot move to another vendor without large switching costs. The lock-in effect is well-known in the software industry since the customer may become dependent on proprietary data formats or interfaces, and high switching costs might prevent the change to another product (Ruttan 2001).

Avoiding the "lock-in" effect has been proposed by some authors as an important motivation for using open source. According to this view, users of FOSS would have the capacity to switch between different vendors and thus reduce their dependence on a single manufacturer. Thus, a Linux user can choose between hardware from different manufacturers, where an IBM mainframe user is constrained in his future choices. However, like all technology products, FOSS is also affected by path dependence, mostly by *increasing returns*. When a user chooses Linux over FreeBSD, he reinforces the notion that Linux is the best choice for a FOSS operating system. Considering that many FOSS products rely on community support for helping novice users and producing documentation, path dependence is very relevant to FOSS. In the long run, increasing returns encourage a single FOSS solution to be dominant in each market segment. By contrast, the existence of many Linux distributions harms its expansion at the desktop. Non-technical users are discouraged by the alternatives and fear for their long-term endurance. We consider that the convergence to a single distribution is a necessary condition for Linux to reach a significant market share of the desktop users.

Disruptive potential. An important factor in the adoption of a FOSS is its disruptive potential, considered as its capacity to provide a novel functionality that can have long-term consequences. Such is the case of Linux. Before Linux emerged as a viable alternative, most Intel-based computers ran Windows, and each major UNIX workstation builder (such as IBM, DEC, SUN, HP and SGI) had its own line of CPUs. However, these companies had a hard time competing with Intel, as the scale of investments each successive line of CPUs increased. Maintaining Moore's Law is expensive. So, the availability of a UNIX-like operating system running on Intel chips caused the demise of non-Intel alternatives such as IRIX and Tru64 Unix, the transformation of Solaris into a FOSS, and the reduction of the market for HP-UX and AIX. Even IBM, which still maintains its AIX operating system, is making substantive investments on Linux (Samuelson 2006). Thus, Linux was a disruptive technology that ultimately changed the whole UNIX-like scene.

Documentation. Another significant factor that limits FOSS large-scale adoption is user documentation. Users have different learning processes, and thus it is necessary to produce different types of manual and tutorials to help them. In most cases, producing such documentation is beyond the capacities of FOSS teams. That is one of the reasons why FOSS is more popular with computer-savvy experts that with the non-technical users.

Community support. Having a strong support from an established community helps to decrease the path dependence effect of proprietary software on users. As an example, the PROJ4 library has an active community that helps to support its use by different software products.

14.3 The Current FOSS4G Scene

We now consider how the factors for adoption for FOSS relate to the current FOSS4G scene, taking as our reference the works of Steiniger and Bocher (2009) and Ramsey (2007). Based on these reviews, we can divide the FOSS4G scene into five clusters:

(a) *Java-based visualization and analysis software:* programs written mostly in Java that enable the visualization of spatial data and provide functions for analysis of such data. They include uDig, gvSIG, KOSMO, and OpenJUMP and its derivatives. These software also provide functions for handling vector data.
(b) *C and C++ -based visualization and analysis software:* programs for visualization and analysis of spatial data using the C and C++ programming languages. These include QGIS, TerraView, and MapWindow. As in case (a) above, these software are capable of handling vector data.
(c) *Image and Raster data analysis software:* programs for analysis of (mostly) raster data, such as images and digital terrain models. These include OSSIM, GRASS, SAGA, SPRING, and ILWIS. GRASS and SPRING also provides significant capabilities for working with vector data.
(d) *Database interfaces:* APIs that provide support for access to object-relational database systems to handle vector data. The best example is PostGIS (Santilli et al. 2007). Other example is the TerraLib library, designed for handling large-scale environmental data. It can access PostGIS and other DBMS's and also allows access and storage to raster data (Câmara et al. 2008). Other notable efforts include GDMS[1] and Hybernate spatial.[2]
(e) *Supporting libraries:* these libraries support functions which are needed by any GIS, and are used by many of the current front-end implementations. They include GDAL/OGR, PROJ4, GEOS, JTS (Java Topology Suite) and Geotools.

There are many additional differences between the above-mentioned software. Some support the OGC standards, others are file-based, and the analysis functionality varies greatly. In terms of visualization, most provide similar capabilities. For vector data analysis, no FOSS4G package provides an equivalent quantity of functions as those available in the

1 http://trac.orbisgis.org/t/wiki/devs/GDMS
2 http://www.hibernatespatial.org

leading proprietary GIS (ESRI's ArcGIS). For raster data analysis, the functionalities of FOSS4G products approach those of proprietary software, although in some niche areas (digital photogrammetry, object-based image analysis) proprietary software still has the lead.

The lack of accurate surveys makes it difficult to assess the impact of FOSS4G on the market. We have to rely on indirect assessments, such as the difference in user attendance in FOSS4G and ESRI's conferences, and the small number of large-scale applications that use FOSS4G. Such scant evidence suggests that none of the available FOSS4G products has reached a significant proportion of desktop users, which continue to rely on proprietary software. So far, no desktop GIS has shown potential for breaking the 'lock-in' effect which proprietary software products have on users. This is a direct consequence of the fragmentation of the community. To understand more, we need to consider both the achievements and challenges to FOSS4G, as follows.

14.3.1 Achievements of the FOSS4G Community

The most important achievement of the community is intangible: FOSS4G exists and is being used worldwide for real-world applications. Many case studies have been presented at the FOSS4G Conference and similar venues. These cases show that the skills required to build a FOSS4G application are similar to those required to use proprietary software.[3] In general, the TCO (total cost of ownership) of a FOSS4G application is significantly less than that of a proprietary application. Personnel salaries and hardware costs are similar, and software costs are much reduced.

A large part of the credit for the success of FOSS4G product is due to the PostGIS project and the activities of the GRASS user community. The GRASS user community includes users from all parts of the world, that actively participate in mailing lists and user conferences.[4] As for PostGIS, we concur with Ramsey (2007): "*The strength of PostGIS is that it has become the standard spatial database backend for all the other open source GIS tools.*" Indeed, programs such as uDig, gvSIG, KOSMO, OpenJUMP, and TerraLib rely on PostGIS as their database support.

PostGIS and the underlying PostgreSQL DBMS have proven capabilities of support industry-grade applications. Take for example TerraAmazon, Brazil's national database for monitoring deforestation in Amazonia.

---

[3] For some metrics on the subject, please see the discussion http://lists.osgeo.org/pipermail/conference_dev/2010-March/001029.html

[4] See for example http://gisws.media.osaka-cu.ac.jp/grass04/

TerraAmazon is based on TerraLib that in turn uses PostGIS with added support for raster storage in PostgreSQL (Freitas et al. 2007). The database currently stores more than three million complex polygons and 600 GB of full resolution satellite images, using the TerraLib pyramidal resolution schema.

Another area of achievement is that of application libraries. Products such as GDAL/OGR, PROJ4 and GEOS are being used by most of the FOSGIS community, due to their stability and quality. These libraries are well-designed and solve important issues in GIS design, namely: interface to data formats (GDAL), cartographic projections (PROJ4) and topological predicates and operations (GEOS and JTS).

14.3.2 Challenges to the FOSGIS Community

The authors consider the division between the 'C/C++', 'Java', and 'raster' clusters as the major impediment for the growth of FOSS4G solutions in the market place. This fragmentation is a good example of the dreaded forking pattern which keeps users away from FOSS. In an ideal but unlikely situation, if all FOSS4G developers would agree to use a single environment, the community could build a system that would surpass the current proprietary software in quality and functionality. As the situation stands, there are too many systems offering similar functions, whereas none has the power to become dominant.

Modularity. Most desktop GIS are monolithic applications, where the visualization actions are closely tied to the handling of interface events. The design choices for GUI design and event handling are hard to share across a community of programmers. There are many incompatible options for GUI design (Qt, GTK+, SWT, SWING) and there are plausible reasons for choosing any of them. Once a GUI base library is chosen, the desktop GIS code becomes hard-wired on that library and is difficult for outsiders to make effective contributions to the software. Because of the hard-wired links between their interface and the underlying GUI library, desktop GIS applications tend to be monolithic and non-modular.

Leadership. Good programmers are notoriously strong-minded and respect only those they view as superior to them. As Fitzgerald (2004) observes: *"to lead an OSS project really successfully, the leader needs to be a 'code god' who inspires others and whose ability and authority is beyond discussion."* Most of the current FOSS4G is being developed independently by like-minded teams, who had limited contact with their peers. While there are recognized programmers in the community, there is no established leadership in the sense of Linus Torvalds and Richard

Stallman. The project leaders of the different desktop GIS products have thus little incentive for merging their activities with those of similar initiatives. The main gathering of the FOSS4G community (the annual FOSS4G Conference) was established after most of the desktop GIS projects were already underway. The FOSS4G Conference serves as an important contact point for members of the community, but so far it has not promoted any convergence of FOSGIS technology. The authors hope that the OSGeo organization (founded in 2006) can act as a catalyst. In an ideal world, OSGeo could support a large scale communitarian effort that would produce a FOSS4G product that could outperform today's proprietary solutions.

Standards and shared conceptualization. Most desktop GIS look similar, since they share a common paradigm for their interface (that of the proprietary software ArcView). Most of them also share the OGC standard interfaces for web services and vector data. However, the OGC effort on GIS standardization is commendable but incomplete. There has been significant progress in OGC for vector data storage, database query, and web services. However, there are no mature standards for data analysis, image processing, map algebra, digital terrain modelling, and spatial statistics.[5] The lack of shared conceptualization for analysis and processing functions has led to incompatible products. Each GIS development team has developed his own implementation of these functions, leading to further incompatibility between the products. Applications such as GRASS, SAGA, SPRING, and ILWIS use many different concepts. Indeed, proprietary image processing software such as ENVI, ERDAS and PCI also have significant differences. As a result, each software has a slow learning curve which creates a lock-in effect on its users. By contrast, the spatial statistics community that uses the R software (Ihaka and Gentleman 1996) has decided to join forces and develop a single product (called R-spatial) that provides most of the functionality they need (Bivand et al. 2008). A further gray area concerns storage of raster data in DBMS. Although ORACLE already provides support for storing raster data in its proprietary products, there is much debate on the FOSS4G community about the advantages and problems of storing raster data in a free spatial DBMS such as PostGIS. This is another area which would benefit from a standard that could be followed by the community.

Code stability and avoidance of forking. The large number of FOSS4G products indicates that designing a desktop GIS is relatively easy, within the

[5] Such lack of standards has motivated many developers to come up with new products, such as OGRS team at IRSTV Nantes (France) and the TerraLib team at INPE (Brazil).

reach of a small team of programmers. This is bad news. It is too easy to start a new project, and this reduces the motivation for adoption a solution developed elsewhere. Some FOSS4G projects, such as JUMP and GeoTools, have gone through phases of forking and merging with other projects. This is negative for code stability and tends to restrain user adoption. Part of the problem has to do with language choice. Currently, most GIS programmers have adopted Java as their preferred development environment. However, many existing products (such as GRASS, SPRING, TerraView, ILWIS) have lots of code already developed and tested using C or C++. This causes an impasse when trying to expand a Java-based visualization and vector handling software into a more complete GIS. The solution here is to develop stable software libraries that could provide all the functionalities of products such as GRASS or SPRING and that could be used by different products. The PostGIS, GDAL, PROJ4 and GEOS libraries are good examples of code stability and continuous improvement that need to be complemented and extended.

Innovation. To stay ahead of the market, companies have to innovate continuously. As Andrew Grove of Intel said, *"only the paranoid survive"*. The GIS market is currently full of activities related to mobile devices and to web services. However, FOSS4G products tend to be conservative and focused on the desktop, based on the map metaphor. By contrast, the commercial vendors (especially ESRI and Google) are investing heavily in mobile GIS applications. Considering the money commercial companies spend on innovation, FOSS4G developers would need to be united behind a small number of innovative products to be able to compete.

Disruptive potential. PostGIS is the FOSS4G product with greatest potential to be a disruptive technology and to change the GIS market. Should PostGIS evolve to support different types of geospatial data needed by emerging GIS applications, such as images, DTMSs, geo-sensor data, and location-based services, it could provide the support FOS GIS needs to provide strong competition to proprietary software. We will return to this issue in the next section. In short, while PostGIS has a good potential to be an important player in the user GIS market, the other desktop GIS lack the momentum to become dominant, and the raster data processing solutions are incompatible and hard to combine. While this split persists, the leading commercial companies continue to innovate and to keep or increase their market share.

14.4 GIS in the Twenty-First Century: What Can Become of FOSS4G?

The preceding sections tried to show why currently there is no good candidate for a Geo-Linux. While PostGIS has a good momentum to provide the services needed by the different types of geospatial applications, there is no dominant solution in the desktop, nor there are emerging applications for mobile devices and web services. Does that mean that there will never be a Geo-Linux? As the baseball manager Yogi Berra said, *"The future ain't what it used to be"*. If the FOSS4G community is to design products that reach a significant market share, it needs to learn the lessons from Linux and Apache. We need products with disruptive potential, not limited to reproducing designs from proprietary products, providing truly innovative capabilities. In this perspective, this section examines two areas which have the potential to bring forth disruptive applications: *Web services and data-intensive GIS*.

Consider Web services. In the early 2000s, the potential of the Web-enabled spatial applications became clear to the GIS community. Their reaction was build software for data visualization and browsing on the Web (such as MapServer) and to establish a set of standards, including WCS, WMS and WFS from the Open Geospatial Consortium (OGC). The abstractions encapsulated in OGC's Web standards deal mostly with a non-cooperative environment. This approach underestimates the Web's potential for innovation and considers it just a medium for visualization of remote information. Using OGC's standards, users have access to information produced by others, mostly for visualization. The user is thus a passive consumer of information produced elsewhere (Câmara et al. 2009). Using social networks in the Internet, the FOSS4G community could build collaborative systems that go beyond the simple OGC abstractions and support cooperation and interaction. Although there are some exceptions, such as Open Street Map, most FOSS4G applications are conservative in their use of the Web's potential.

The Spatial Web GIS is more and more being dominated by Google Maps and Google Earth. Google developers come from a different mindset than the map-centred GIS community. They recognized the value of supplying the base location layer (vectors and images) as a seamless data set, and of providing APIs for extensibility. The Google Maps API allows programmers to embed Google Maps in their web pages and to add content to the base layers. The KML format and the Google Earth API allow the user to add maps, images, routes, and other spatial documents to Google Earth, thus easing the task of creating a Web GIS server. Such flexibility and performance comes at a price: users must comply with the IP restrictions set by Google. From a FOSS4G viewpoint, Google's potential for dominance of the Spatial Web is dismaying. Should it happen, the control of the Spatial Web by a proprietary solution will limit both the development of FOSS4G and reduce the incentive for sharing data, either voluntary or government produced. It is time the FOSS4G community learns the "Google lesson": *Those that control the data control the future of the Spatial Web.* Google Maps and Google Earth show what can be done when only one company has access to key spatial data, such as street networks and remote sensing images.

The limitations of RDBMS and SQL for handling geospatial data have been recognized long ago (Egenhofer 1992). Nevertheless, object-relational databases have been adopted by the GIS community as the primary way to organize collections of spatial data. Using PostGIS or its proprietary alternatives, GIS users have built huge datasets, each modelled using a different conceptual schema. When these datasets were created in the early to mid 2000s, the prevailing business model was to have public data sold on a cost-recovery basis. This business model is changing and national mapping agencies are moving towards an open access model. In April 2010, the UK Ordnance Survey launched a service offering free and unrestricted access to most of its map data. Similarly, the US Geological Survey has released its complete LANDSAT imagery archive. Many other mapping agencies and image providers have released their datasets or are planning to do so. When achieved, full open access to spatial data will bring a complete change on the way GIS technology is used and will provide a unique opportunity for FOSS4G: *the emergence of data-intensive GIS.*

We define *data-intensive GIS* as a software environment capable of handling distributed geospatial data sets, each with its own archival policy. These datasets include spatial databases in OGC format, WCS, WFS and WMS servers, and semi-structured data in different formats (flat files, indexed collections, relational tables). A data-intensive GIS application would be able to use these data sets in a coherent way combining different sources and making appropriate semantic inferences whenever needed. A data-intensive GIS would use techniques that go beyond what the current extensions of SQL allow. For example, such techniques would allow multiple passes over the same data to compute spatial algorithms, a task with is cumbersome to perform using SQL queries. They would also combine different types of spatio-temporal data in a way not possible for SQL and its extensions.

Consider a situation where all government-produced census data, streets and routes, and satellite imagery would be fully accessible on the Web, based on copy-left licenses such as GPL or Creative Commons. The

FOSGIS community could design, build and maintain applications that would handle distributed spatial data. They would use powerful inference techniques to make sense of the data they collect. These data-intensive GIS application could disrupt the current market and change the way we deal with geospatial information.

If the above proposal seems far-fetched, consider that similar techniques are already available. Again, Google is the leading innovator. Google's Map-Reduce technique has shown to be efficient for processing multi-terabyte datasets in parallel (Dean and Ghemawat 2010). A Map-Reduce job has two steps. First, a master node splits the input dataset into independent chunks and passes each of them to a secondary node for processing. When the master node receives the results, sorts them, and passes which are processed by worker nodes in a completely parallel manner. The master node then takes the answers to all the sub-problems and combines them to answer the problem it was originally trying to solve. Typically both the input and the output of the job are stored in a file-system. More than 10,000 distinct programs have been implemented using Map-Reduce at Google, including algorithms for large-scale graph processing, text processing, machine learning, and statistical machine translation (Dean and Ghemawat 2010). Even the defendants of RDBMS recognize that there are processing tasks that are well-suited for Map-Reduce (Stonebraker et al. 2010). The Map-Reduce paradigm is suited for investigation as one of the alternatives to SQL queries for handling geospatial data.

In resume, the emergence of open access policies for geospatial data will provide an opportunity for the FOSS4G community to develop innovative and disruptive applications. To succeed, GIS developers will have to overcome a number of challenges. These include modelling the semantics of spatial concepts, understanding change in space and time, and developing information extraction methods for massive data sources (Câmara et al. 2009). The success of information technologies such as Google's shows that these challenges are hard, but not impossible to take on. Should the FOSS4G community succeed in producing innovative solutions for data-intensive GIS, it is likely that a Geo-Linux will finally emerge.

14.5 Concluding Remarks

This paper discusses the current situation of FOSS4G. We recognize that the FOSS4G community has already achieved relevant results. There are many good desktop FOSS4G available, which support the basic tasks of retrieval and visualization. The various OS raster data processing software have lots of functions. However, FOSS4G products still have a small market share. In this paper, we try to explain why this happens. Based on the properties of open source software in general, we argue that there is excessive fragmentation on the FOSS4G scene. Also, current FOSS4G are not modular enough to support large development teams. Furthermore, there is a lack of common standards for many important GIS applications, such as map algebra, spatial statistics and image processing, leading to similar but incompatible products. More important, no current FOSS4G has a disruptive potential that could unbalance the market. We also indicate that there are structural reasons in GIS design that contribute to produce this status quo.

Then, the authors examine two areas that are relevant for the future of GIS: Web services and data-intensive GIS. We argue that the trend towards open access geospatial data brings an opportunity for FOSS4G. We could develop disruptive GIS that could support data-intensive applications, combing data from diverse sources and allowing the user to overcome semantic barriers. If the FOSS4G community succeeds in developing such an innovative product, we could finally have a Geo-Linux. The enormous potential of a dedicated community would be at last realized.

Acknowledgments

The first author thanks Erwan Bocher and his team at Nantes for organizing the OGRS 2009 conference and inviting him to deliver a keynote speech, which led to this paper. The authors also thank the Terralib development team at INPE, especially Antonio Miguel Monteiro, Gilberto Ribeiro, and Karine Ferreira for many hours of inspired discussion. Gilberto Camara's work is partially funded by CNPq (grant PQ 550250/2005-0) and FAPESP (grant 04/11012-0).

References

Bivand R, Pebesma E, Gómez-Rubio V (2008) Applied spatial data analysis with R. Springer, New York

Brooks F (1987) No silver bullet: essence and accidents of software engineering. IEEE Comput 20:10–19

Câmara G, Fonseca F (2007) Information policies and open source software in developing countries. J Am Soc Inf Sci Technol 58:121–132

Câmara G, Onsrud H (2004) Open source GIS software: myths and realities. In: Esanu JM, Uhlir PF (eds) Open access and the public domain in digital data and information for science: proceedings of an international symposium. The National Academies Press, Washington, pp 127–133

Câmara G, Vinhas L, Ferreira K, Queiroz G, Souza RCM, Monteiro AM, Carvalho MT, Casanova MA, Freitas UM (2008) TerraLib: an open-source GIS library for large-scale environmental and socio-economic applications. In: Hall B, Leahy M (eds) Open source approaches to spatial data handling. Springer, Berlin, pp 247–270

Câmara G, Vinhas L, Davis C, Fonseca F, Carneiro T (2009) Geographical information engineering in the twenty first century. In: Navratil G (ed) Research trends in geographic information science. Springer, Berlin, pp 203–218

Chambers JM (1998) Programming with data. Springer, New York

David P (2007) Path dependence: a foundational concept for historical social science. Cliometrica 1:91–114

Dean J, Ghemawat S (2010) MapReduce: a flexible data processing tool. Commun ACM 53:72–77

Egenhofer M (1992) Why not SQL! Int J Geographical Inf Syst 6:71–85

Fitzgerald B (2004) A critical look at open source. IEEE Comput 37:92–94

Freitas U, Ribeiro V, Queiroz GR, Petinatti M, Abreu E (2007) The Amazon deforestation monitoring system: a large environmental database developed on terraLib and post-greSQL. OSGEO J 3:70–75

Ghosh RA, Rudiger Glott, Kreiger B, Gregario Robles (2002) The Free/Libre and open source software survey and study – FLOSS Final Report. International Institute of Infonomics, University of Maastricht, Maastricht, The Netherlands

Hall G, Leahy M (2008) Open source approaches in spatial data handling. Springer, Berlin

Ihaka R, Gentleman R (1996) R: a language for data analysis and graphics. J Comput Graphical Stat 5:299–314

Munga N, Fogwill T, Williams Q (2009) The adoption of open source software in business models: a Red Hat and IBM case study. In: Proceedings of the 2009 annual research conference of the South African institute of computer scientists and information technologists. ACM, Emfuleni, Vanderbijlpark, 2009

Nagy D, Yassin AM, Bhattacherjee A (2010) Organizational adoption of open source software: barriers and remedies. Commun ACM 53:148–151

O'Reilly T (1999) Lesson from Open Source Software Development. Commun ACM 42(4):33–37

Page S (2006) Path dependence. Q J Political Sci 1:87–115

Ramsey P (2007) The state of open source GIS. Refractions Research, Victoria

Ruttan V (2001) Technology, growth and development. Oxford University Press, New York

Samuelson P (2006) IBM's pragmatic embrace of open source. Commun ACM 49:21–25

Santilli S, Leslie M, Hodgson C, Ramsey P, Lounsbury J, Blasby D (2007) PostGIS manual – version 1.3.2. Refractions Research, Victoria
Steiniger S, Bocher E (2009) An overview on current free and open source desktop GIS developments. Int J Geographic Inf Sci 23:1345–1370
Steiniger S, Hay GJ (2009) Free and open source geographic information tools for landscape ecology. Ecol Info 4:183–195
Stonebraker M, Abadi D, DeWitt DJ, Madden S, Paulson E, Pavlo A, Rasin A (2010) MapReduce and parallel DBMSs: friends or foes? Commun ACM 53:64–71
Weber S (2004) The success of open source. Harvard University Press, Cambridge
Yu L, Schach SR, Chen K, Heller GZ, Offutt J (2006) Maintainability of the kernels of open-source operating systems: a comparison of Linux with FreeBSD, NetBSD, and OpenBSD. J Sys Software 79:807–815

[illegible]

Chapter 15 Free and Open Source GIS Software for Building a Spatial Data Infrastructure

Stefan Steiniger, Andrew J. S. Hunter

Department of Geomatics Engineering, University of Calgary, 2500 University Drive NW, Calgary, AB T2N 1N4, Canada,
{ssteinig, ahunter}@ucalgary.ca

Abstract. The implementation of Spatial Data Infrastructures (SDIs) for governments and companies is a task that has gained ample attention in recent years. Different categories of spatial software such as desktop GIS, server GIS, web map servers, spatial database management systems, web map development toolkits, etc., are required to realize the software components of an SDI. We catalogue a (selected) variety of free and open source projects that develop and maintain spatial software that fit within these categories. Our analysis reveals that for all categories of software used in SDIs a free software product is available. This enables adopters to implement an SDI on a limited financial budget, and allows the distribution of a proven SDI architecture without legal constraints. Our software evaluation shows that free and open source solutions support a wide range of industry standards that ease interoperability between SDI components.

15.1 Introduction

Over the past decade government organizations at provincial, regional, national and international levels, as well as non-governmental organizations and large companies, have been working on the implementation of Spatial Data Infrastructures (SDIs) – with the objective of easing access to spatial data for a wide range of users (Rajabifard and Williamson 2001; Crompvoets et al. 2004). For instance, in Canada alone 93 new spatial portals and systems for spatial data dissemination have been built, and 251

E. Bocher and M. Neteler (eds.), *Geospatial Free and Open Source Software in the 21st Century*, Lecture Notes in Geoinformation and Cartography,
DOI 10.1007/978-3-642-10595-1_15,

new datasets published using GeoConnections[1] funding between the years 2005 and 2009 (GeoConnections 2009). Building an SDI requires a range of software which may depend on the SDI participants' perspective: geospatial data provider, geospatial service provider, or geospatial data user. To support these ongoing implementation efforts we will (i) give an overview of the categories of software required, and (ii) list free and open source projects that develop and maintain such software. In particular, with this summary of free software projects we wish to aid persons charged with building and maintaining SDIs by pointing to suitable candidates from the free software ecosystem that we believe are useful for the implementation of an SDI. We have chosen to focus on Free and Open Source Software (FOSS) as we believe that these solutions will further the implementation of SDIs in developing countries, government and industry, because of their ease of distribution, and because of their conformance to interoperability standards.

To our knowledge there have only been two earlier publications on the topic of FOSS for SDIs; Reid and Martin (2001) and May et al. (2005). However, both were published several years ago, and there has since been considerable development in the free software community – as such we believe an update is warranted.

We will start the overview by introducing the components of an SDI, summarize the technical standards used to distribute and process spatial data, and identify the spatial software categories that are required to build an SDI. Subsequently we will introduce into free software terminology and briefly review FOSS literature that discusses free spatial software. Finally, we will catalogue, non-exhaustively, free spatial software projects that develop software that can be used in SDIs. We note, that we have decided to only list the projects, without description, due to the large number of projects – and space constraints.

15.2 Spatial Data Infrastructures

Before mass use of the Internet and its technologies, and still today to a large extent, spatial data for a particular location have been stored in different physical locations, and often using different standards or formats. This makes it difficult for a potential user to access and utilize the data. Smits (2003)

[1] GeoConnections is a Canadian program that aims to make location-based data and technologies broadly accessible to decision makers (see http://www.geoconnections.ca).

has described this situation as "[...] islands of data of different standards and quality...". Potential users of these "islands of data" might be organizations that cannot afford to acquire data on their own, or need data from outside their jurisdictions (Rajabifard and Williamson 2001). Hence, the primary objective for creating an SDI is to provide these potential users access to (spatial) data stored somewhere else (Clinton 1993), i.e., the sharing of data, to save resources, time and effort, by avoiding the duplication of effort required to acquire and maintain the data (Rajabifard and Williamson 2001).

Besides the technical components that connect data 'islands' via the Internet, other aspects such as licensing agreements, data transfer standards, and data access policies must also be put in place to ensure consistent and reliable access. Consequently, a spatial data infrastructure is not only a technical facility, but a complete framework that includes political, technical, business and social aspects. From the diverse literature on SDIs (Rajabifard and Williamson 2001; Crompvoets et al. 2004; Smits 2003; Granell et al. 2009; Kuhn 2005; GSDI 2009; Smits et al. 2002) we found it useful to distinguish the following five components:

- Spatial Data (or spatial information)
- Technologies, i.e., hardware and software
- Laws and Policies
- People, i.e.: data providers, service providers, users, and
- Standards for data acquisition, representation and transfer

A definition for SDI given by Kuhn (2005) includes more or less all of these components (emphasis from the original text):

> An SDI is a **coordinated** series of **agreements** on **technology** standards, **institutional** arrangements, and **policies** that enable the **discovery** and **use** of geospatial information by users and for purposes **other** than those it was created for.

15.2.1 GIS Software Needs in SDIs

A Spatial Data Infrastructure should enable the discovery and delivery of spatial data from a data repository,[2] ideally via one or more web services. Additionally, it is often desirable that the data provider is able to (remotely) create and update spatial data stored in a repository. Hence, the

[2] For our purposes a data repository is a collection of spatial datasets, that may be stored in one or more spatial databases, on one or more physical servers, i.e., a data warehouse. We assume that the repository is integrated and non-volatile.

basic software components of an SDI consists of (i) a software client that can display, query, and analyse spatial data, (ii) a catalogue service for the discovery, browsing, and querying of metadata or spatial services, spatial datasets and other resources, (iii) a spatial data service that enables the delivery of the data via the Internet, and/or processing services such as datum and projection transformations, (iv) a (spatial) data repository, and (v) GIS software (client or desktop) that permits the creation and maintenance of data (Fig. 15.1). The catalogue service (item ii above) functions in a similar manner to analogue directories such as the Yellow Pages. The extent of processing services (iii above) is somewhat unlimited, but transformation services are generally fundamental to a distributed system.

15.2.2 Important Technical Standards for Spatial Data Distribution

Several technical standards defined by the Open Geospatial Consortium (OGC, opengeospatial.org) and the International Organization for Standardization (ISO 19xxx series, iso.org) play an important role in the dissemination and processing of spatial data. In general, these standards describe communication protocols between data servers, servers that provide spatial services, and client software, which request and display spatial data. In addition, they define a format for the transmission of spatial data. We identify the following OGC standards that are required to build an SDI:

- OGC data delivery standards: Web Mapping Service (WMS), Web Feature Service (WFS) and its transactional equivalent (WFS-T), and the Web Coverage Service (WCS)
- OGC data format standards: Simple Feature Standard (SFS), Geography Markup Language (GML), Keyhole Markup Language (KML)
- OGC data search standards: Catalogue Service (CSW), Gazetteer Service (WFS-G) and
- Other OGC standards: Web Processing Service (WPS), Coordinate Transformation Service (CTS), Web Terrain Service (WTS), Styled Layer Descriptor (SLD), Symbology Encoding (SE), Web Map Context (WMC)

The ISO Standard 19115 (Geographic Information – Metadata) is also important as it defines a schema for describing spatial data, and ISO 19119 (Geographic Information – Services) defines how spatial data and services should be described so that they are able to be searched by catalogue services (such as OGC's Catalogue Service). Specifications developed by the World Wide Web Consortium (W3C) for data dissemination such as HTML, XML, SVG, SOAP, WSDL, etc., should also be considered (see Chap. 7 of the SDI Cookbook (GSDI 2009)). Further information on applicable and implemented standards can be found in (GSDI 2009; Smits et al. 2002; Lupp 2008).

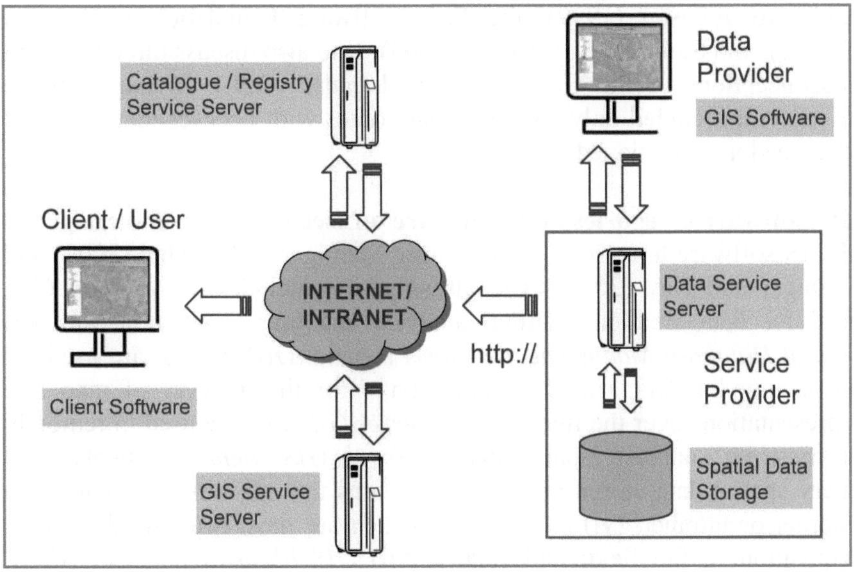

Fig. 15.1. Software needs of an SDI

15.3 Free and Open Source Spatial Software

Defining terms: free and open source – Software that is called '*Free Software*' or '*Free and Open Source Software*' (FOSS) is covered by a specific type of software license. This license needs to grant the software user at least the following four freedoms: (1) the freedom to run the program, for any purpose (e.g., no matter if for educational or business purposes) (2) the freedom to study and adapt the program for one's own needs, (3) the freedom to redistribute the software, and (4) the freedom to improve the software and to release these improvements to the public (see www.gnu.org/philosophy/free-sw.html). In this sense "*free*" takes on concepts of the public domain such that there are no rights to exclude. Hence, software that is (i) gratis, i.e., free-of-cost, and (ii) permits viewing of the source code does not automatically qualify as "free and open source" software. If the user's rights include the right of independent use, of modification, and of free distribution – only then – can the software be called "free" and "open source" software. With respect to this definition a distinction between the categories of free/gratis vs. commercial/paid for software is not entirely correct, since the freedoms outlined above do not prohibit a developer from attaching a cost to the distribution and support of "free software". For a discussion of free software licenses we refer the reader to Johnson (1999), the Free Software Foundation web pages (fsf.org), and Steiniger and Bocher (2009), who also discuss three common misconceptions of the General Public License (GPL). In our software review, we included only software that comes with licenses that grant the four freedoms mentioned above.

GIS software categories and literature on free GIS – Several categories of GIS software have been identified by Steiniger and Weibel (2009) and Steiniger and Hay (2009). The classification includes: (i) *desktop GIS*, used for data creation, editing, analysis and map generation; and (ii) *Spatial Database Management Systems* (Spatial DBMS) that are used for the storage of data; (iii) *web map server* for the delivery of map like representations over the Internet; (iv) *server GIS*, that are used to remotely process, e.g., analyse, spatial data; (v) *web GIS clients*, to display and query spatial data stored at remote locations that are only accessible via Internet or intranet; (vi) *mobile GIS*, which are most often used for data acquisition in the field; and finally (vii) *GIS libraries and extensions*, which provide special functions that can enhance standard (desktop) GIS capabilities, or be used to build customized GIS applications, including web mapping applications.

Reviews of Free and Open Source Software for GIS (FOSS4GIS) and its utility have become more common over the last three years as several free software solutions have matured to the point that they are able to replace proprietary software. An early assessment of FOSS4GIS focusing on SDIs was completed by Reid and Martin (2001), who concluded in 2001 that there was great potential for the use of FOSS4GIS solutions that incorporated SDI components. FOSS4GIS software reviews not related to SDIs but rather environmental GIS applications have been completed by Jolma et al. (2008) and Steiniger and Hay (2009). Ramsey's (2007) general review of FOSS4GIS projects, Sanz-Salinas and Montesinos-Lajara (2009) summary of major FOSS4GIS projects, and Steiniger and Bocher's (2009) as well as Donnelly's (2010) recent evaluation of FOSS4GIS desktop projects may also be of interest to SDI implementers.

15.4 Free and Open Source Spatial Software for SDIs

In the previous sections we have outlined the types of GIS software and the major software components that exist in a general SDI architecture. For the discussion below we have combined those categories and arranged the discussion as follows: (4.1) SDI server software: (4.1-A) web map servers (4.1-B) server GIS; (4.2) SDI spatial data storage: Spatial Data Base Management Systems (Spatial DBMS); (4.3) SDI registry/catalogue and metadata software; (4.4) SDI client software: (4.4-A) desktop GIS clients, (4.4-B) web-GIS development toolkits (Fig. 15.2).

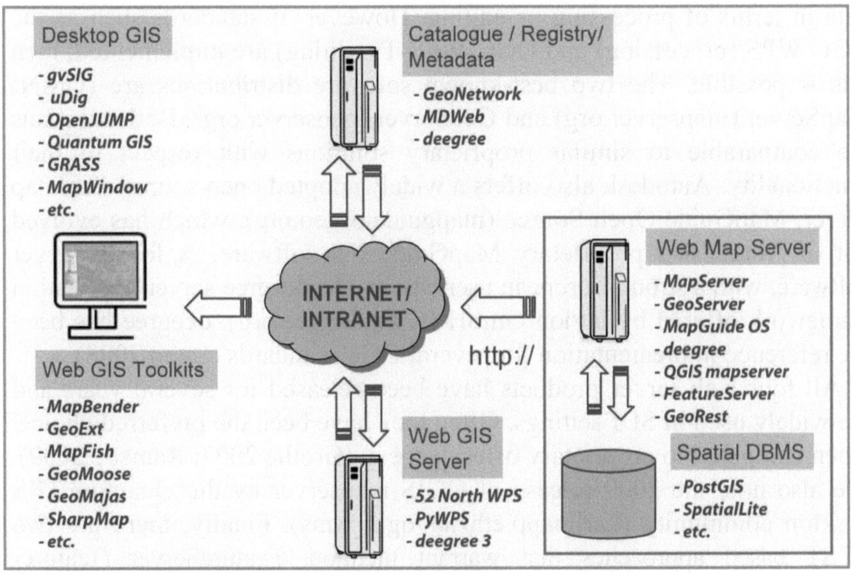

Fig. 15.2. FOSS4GIS software that can be used in an SDI

Our review below aims to focus only on the more established projects. For some categories we will also mention software under development if we believe the project could attract a wider user group. Due to page limits we will only list the software projects and not describe them in detail – and instead refer the interested reader to the related project web pages and Sanz-Salinas and Montesinos-Lajara (2009). However, we try to give a general assessment for the software categories with respect to software advancement and use in comparison to proprietary software, and list in Table 15.1 the supported OGC/ISO standards for selected software.

15.4.1 SDI Server Software

In the following we distinguish between two different categories of server software, web map and web-GIS servers, owing to their different usage (Steiniger and Weibel 2009; Steiniger and Hay 2009).

A – Web map servers for serving data and images – Web mapping services, i.e., servers, usually serve spatial data based on three OGC standards: WMS (or ISO 19128: 2005, Geographic Information – Web Map Services) for the display of maps as image, WFS for vector data, and WCS for raster data. Web map servers rarely allow interaction with the data in terms of processing or editing. However, if standards such as the OGC WPS (processing) and OGC WFS-T (editing) are implemented, then this is possible. The two best-known software distributions are (UMN) MapServer (mapserver.org) and GeoServer (geoserver.org). Both solutions are comparable to similar proprietary solutions with respect to their functionality. Autodesk also offers a widely adopted open source web map server, MapGuide Open Source (mapguide.osgeo.org), which has evolved out of Autodesk's proprietary MapGuide 6.5 software. A fourth server software, with a strong European user base, is the deegree server/application framework offered by lat/lon GmbH (wiki.deegree.org). deegree has been the reference implementation for several OGC standards (Lupp 2008).

All four web server products have been released for several years and are widely used in SDI settings. Often they have been the preferred choice, when compared to proprietary offerings, e.g. (Grothe 2007; Ramsey 2009). We also note the 2009 release of QGIS mapserver by the Quantum GIS desktop community (karlinapp.ethz.ch/qgis_wms). Finally, there are two REST based approaches that warrant mention, FeatureServer (featureserver.org), and GeoRest (code.google.com/p/georest). In a RESTful style service everything is treated as a resource and representations of resources can be uniquely addressed by a universal resource identifier (URI). Both systems offer WFS capabilities and provide various contemporary web formats such as GeoJSON, KML and GeoRSS.

B – Web-GIS servers for data processing – These solutions are distinct from web map servers as they do not provide web mapping services, but rather they expose GIS processing functionality that is typically offered by desktop GIS software – for example data analysis and transformation services. To use the processing functionality the software should support the OGC WPS standard (or ISO 19119: 2005, Geographic Information – Services), which in turn acts as our selection criterion for this listing.

One solution that implements the WPS standard is 52N WPS software by 52°North (52north.org). Spatial analysis, i.e., processing functions, provided by the 52°North software stem from the Sextante project. 52°North also distributes two client plug-ins for the free desktop GIS' uDig and JUMP / OpenJUMP. A further implementation of the WPS standard for server-side use is part of the deegree 3 framework (wiki.deegree.org). The software PyWPS implements the OGC WPS standard (pywps.wald. intevation.org) and enables access to a wide range of analysis functions for the desktop GIS GRASS. Finally, the new ZOO Project (zoo-project.org) also aims to develop software to access and distribute WPS services.

15.4.2 SDI Data Storage Software: Spatial DBMS

Spatial Database Management Systems (DBMS) are usually not developed as a software solution on their own, but rather developers add spatial capabilities to existing free DBMS. The product with the largest user base is probably PostGIS (postgis.refractions.net). PostGIS adds spatial data types and analysis functions to the FOS database PostgreSQL (postgresql. org). PostGIS is currently used in high demand installations, for example by the French National Mapping Agency (IGN France), and it provides the most complete implementation of OGCs Simple Feature Specification (OGC SFS) of any FOSS DBMS solution. Comparisons with proprietary Spatial DBMS, such as Oracle 10g Spatial, show that PostGIS is a comparable alternative when considering functionality, robustness, support and price (Greener 2009).

Two other FOSS DBMS projects include spatial data capabilities. They are the MySQL Spatial extension (mysql.com), which adds spatial support to MySQL, and the SpatiaLite project for SQLite (gaia-gis.it/spatialite). Currently MySQL Spatial provides a basic implementation of OGCs SFS for SQL standard, but query and analysis operations utilize (Minimum) Bounding Rectangles instead of true geometries. Hence, operations have lower precision and may not be useful for everyone (BostonGIS 2009). SpatiaLite realizes conformance to the OGC Simple Features Specification by utilizing the functionality of the GEOS geometry library (trac.osgeo. org/geos).

In addition to the above, there are a number of projects adding spatial support to DBMS, that include (i) Hibernate (hibernatespatial.org), (ii) H2 DBMS (see the websites geoserver.org, orbisgis.org, and hatbox. sourceforge.net), and (iii) INGRES (community.ingres.com/wiki/Ingres Geospatial). Also CouchDB (couchdb.org, Mische 2009) and MongoDb (mongodb.org, Corti 2009) have been used for storing spatial data. Both projects offer a JSON (JavaScript Object Notation) document-oriented database with query and indexing capabilities. However, there are currently no spatial extensions that we are aware of that enable native spatial queries in these 'NoSQL' database implementations.

Table 15.1 Selected OGC standards supported by some of the GIS software mentioned

Software category	Software	Selected supported OGC & ISO standards
Web Map Server	MapServer	WMS, WFS, WCS, SLD, WMC, GML
	GeoServer	WMS, WFS, WFS-T, WCS, SLD, GML, KML
	Deegree 3	WMS, WFS, WFS-T, WFS-G, WCS, WMC, SLD, GML, WPS, CS-W
	Map Guide Open Source	WMS, WFS
	QGIS Mapserver	WMS, SLD
Registry/ MetaData	GeoNetwork	CS-W, ISO 19115:2003, Geographic Information - Metadata
	Deegree 3	see above, ISO 19115, ISO/TS 19139:2007, Geographic Information – Metadata – XML Schema Implementation, ISO 19107: 2003, Geographic Information – Spatial Schema
	CADMEdit,	ISO 19115
	MDweb	ISO 19115, ISO 15836: 2003, Information and Documentation – The Dublin Core Metadata element set)
Desktop GIS[a]	Quantum GIS	WMS, WFS, SFS, GML, KML
	OpenJUMP	WMS, WFS[b], SFS, GML, KML[b], SLD, WFS-T[b], WPS[b]
	gvSIG	WMS, WFS, WFS-G, WCS, GML, KML, CS-W
	uDig	WMS, WFS, WFS-T, SFS, GML, SLD, WPS[b]
	MapWindow	WMS[b], WFS[b]
	GRASS	WMS, WFS, GML, WPS[b]

[a]After writing this article ILWIS Open 3.7 received WMS support; [b]Supported via plug-in.

15.4.3 SDI Registry/Catalogue and Metadata Software

In our search for software falling under the registry/metadata category we found four projects. Three projects provide software to setup a server that functions as "yellow pages" like inventory for spatial data. The first project is GeoNetwork Opensource (geonetwork-opensource.org) funded and used by several large agencies including the Food and Agriculture Organization of the United Nations, the World Food Programme, the United Nations Environment Programme, and the United Nations Office for the Coordination of Human Affairs. MDweb (mdweb-project.org) is a project initiated by research institutes in France and the software is used in several environmental projects and institutions. The deegree framework (wiki.deegree.org) also provides metadata services for the creation, storage, query, retrieval and display of metadata. The fourth software is not a server-based catalogue software, but a metadata editor for desktop computing: CatMDEdit (catmdedit.sourceforge.net). The software allows creation and editing of metadata for several types of GIS file formats and is used by a number of public agencies such as the EUs Statistical Office EuroStat, the National Geographic Institute of Spain and the French National Mapping Agency.

15.4.4 SDI Client Software

A – Desktop GIS clients for data update and analysis – Desktop GIS can be considered the work horse of GIS software. As such, there are a number of FOSS projects that develop and maintain desktop GIS. Recent reviews (Steiniger and Bocher 2009; Steiniger and Hay 2009; Donelly 2010) identify eight mature FOSS desktop GIS. We use the term "mature" to refer to software solutions that offer extensive software functionality, software support, and an international focus driven by both user and developer communities. In order to use desktop GIS in an SDI context the software should provide at least OGC WMS capabilities to ensure that it can act as an SDI client. As not all eight desktop GIS have WMS support our list is reduced to seven: (1) GRASS GIS (grass.osgeo.org), (2) Quantum GIS (qgis.org), (3) uDig (udig.refractions.net), (4) OpenJUMP (openjump.org), (5) MapWindow GIS (mapwindow.org), (6) gvSIG (gvsig.org), and (7) ILWIS Open (52north.org). More details on the capabilities of these software can be found in (Steiniger and Bocher 2009).

Besides these projects some more local or specialized projects exist, e.g., the KOSMO project with a focus on Spanish users (opengis.es). In addition, several newer projects can be found that currently have small user and developer communities, e.g., iGeoDesktop (deegree.org), Orbis GIS (orbisgis.org), and Puzzle GIS (puzzle-gis.codehaus.org).

B – Web-GIS development toolkits for browser-based clients – As an alternative to desktop GIS software, sometimes referred to as a *thick* client, it is becoming more common to make use of a web browser-based *thin* client to display (and query) data. However, to construct a standard web browser for the display of spatial data requires additional functionality on either the client, the server, or both. Several toolkits are available that aid the development of web map pages – or complete spatial data portals.

A simple web map page contains usually only a data viewer with no advanced query or styling functions, although these can be added. To incorporate a simple interactive map the OpenLayers library (openlayers.org) provides an OGC WMS/WFS compatible spatial data viewer. The project OpenScales (openscales.org) provides another option with its Adobe Flex-based viewer. Several other map viewer projects exist or have existed, e.g., ka-Map, Chameleon, ModestMaps and MapBuilder, but several of those projects joined the OpenLayers project (Adams 2009). Hence, there has been an increase in the adoption and development of OpenLayers recently. Two more viewers should be mentioned. The first is SLMapViewer (slmapviewer.codeplex.com), which provides a Microsoft Silverlight interface for MapServer users. The second viewer, MapBender (mapbender.org), is not only a viewer but a web mapping framework that provides functions for managing spatial data services as well, i.e., MapBender could be used to create an SDI portal.

Several development toolkits exist that include not only a viewer (often OpenLayers) but also other tools that help to build web map pages for display and query (or even editing) of spatial data. Projects that provide such frameworks include: GeoMajas (geomajas.org), GeoExt (geoext.org), GeoMoose (geomoose.org), MapFish (mapfish.org), Mapstraction (mapstraction.com), and SharpMap (codeplex.com/SharpMap). If more complex web applications are to be developed, then it is also recommended that developers consider the use of Googles Web Toolkit (gwt.google.com) and Google Gears (gears.google.com). GeoTools (geotools.org) should also be considered to realize server side like GIS functionality and spatial data access.

15.5 Concluding Remarks

Our software review shows that for all categories of GIS software required for the implementation of an SDI, a free and open software product is available. A more detailed examination also reveals that several free products are able to compete with proprietary software, in particular the web map server projects, the spatial DBMS PostGIS, and the data viewer OpenLayers. Indicators for competitiveness in these instances are the number of “real-world installations” and their use in high-demand environments. However, if an organization chooses to implement an SDI using FOSS, then the appropriateness of FOSS projects and software need to be assessed. Key criteria for such an assessment are: (i) functionality, (ii) support of common data distribution standards (e.g., OGC, ISO, CEN standards), (iii) supported platforms (Windows vs. Linux, etc.), (iv) software reliability, (v) options for customization, (vi) distribution costs, (vii) installation and maintenance effort (incl. costs), (viii) training and support options including pricing (Sveen 2008). For a first evaluation of the software it is recommended that adopters make use of GIS distribution packages (for Linux), such as HostGIS (hostgis.com), or Virtual Machine GIS installations, such as GIS VM (gisvm.com) and OSGeo’s Arramagong (live.osgeo.org).

The licenses used by free and open source products ensure a low acquisition cost (typically no cost) for the software itself, and ensure that the software is customizable and adaptable to the SDI context. Hence, and with respect to the latter point, free and open source software often implement a wide set of (OGC) standards, and if not, at least permit the implementation/addition of components that are OGC/ISO standard compliant. The use of open source licenses allows an adopter to deploy SDI components to other locations at no additional cost, e.g., for horizontal scaling of a system. Finally, we note that helpful user and developer communities exist, and that support and maintenance options are offered by various companies, which are similar to those currently offered by proprietary software vendors. In summary, we see that FOSS4GIS software can provide a viable alternative to proprietary products – and we hope that our review helps to assess the utility of FOSS4GIS for SDIs.

Acknowledgments

Parts of this work have been accomplished under a Phase IV GEOIDE project, #03. We are grateful for their funding. We thank the free GIS community, and Erwan Bocher and Markus Neteler for encouraging us to write this overview for the OGRS book.

References

Adams T (2009) There is no alternative to OpenLayers…? Presentation at FOSS4G 2009 conference, Sydney, Australia. Available from http://maps.terrestris.de/dokuwiki/doku.php?id=alternative_openlayers accessed on 28 May 2010

Boston GIS (2009) Cross compare SQL server 2008 spatial, PostgreSQL/PostGIS 1.3-1.4, MySQL 5-6. Available from bostongis.com accessed on 28 May 2010

Clinton B (1993) Coordinating geographic data acquisition and access: the national spatialdata infrastructure. Executive Order 12906. Federal Register, Presidential Documents, 59(71) p 4

Corti P (2009) Using MongoDb to store geographic data. Available from paolocorti.net accessed on 28 May 2010

Crompvoets J, Bregt A, Rajabifard A, Williamson IP (2004) Assessing the worldwide developments of national spatial data clearinghouses. Int J Geograph Inf Sci 18:665–689

Donelly FP (2010) Evaluating open source GIS for libraries. Library Hi Tech 28:131–151

GeoConnections (2009) Status report to the united nations: Canadian geospatial data infrastructure. Natural Resources Canada, Ottawa

Granell C, Gould M, Manso MA (2009) Spatial data infrastructures. In: Karimi HA (ed) Handbook of research on geoinformatics. Hershey, New York, pp 36–41

Greener S (2009) Oracle spatial for PostGIS users – understand, isolate and migrate. Presentation at FOSS4G 2009 Conference, Sydney, AU. Available from http://www.spatialdbadvisor.com/file_download/43/FOSS4G_2009_PostGIS AndOracleSpatial.pdf accessed on 28 May 2010

Grothe M (2007) implementing INSPIRE (principles) – towards an Enterprise SDI for Rijkswaterstaat. Presentation at 110th EuroSDR Meeting, Rotterdam, NL. Available from www.eurosdr.net/meetings/110/presentations/h-5.pdf accessed on 28 May 2010

GSDI (2009) Spatial data infrastructure cookbook. Global Spatial Data Infrastructure Association. Available from www.gsdi.org/gsdicookbookindex accessed on 28 May 2010

Johnson K (1999) Open-source Software Development. Available from http://chinese-school.netfirms.com/computer-article-open-source.html accessed on 28 May 2010

Jolma A, Ames DP, Horning N et al (2008) Free and open source geospatial tools for environmental modelling and management. In: Jakeman AJ et al (eds)

Environmental modelling, software and decision support. Elsevier, Amsterdam, pp 163–180
Kuhn W (2005) Introduction to spatial data infrastructures. Presentation held on March 14 2005. Available from http://www.docstoc.com/docs/2697206/Introduction-to-Spatial–Data-Infrastructures accessed on 28 May 2010
Lupp M (2008) Degree free software. In: Shekhar S, Xiong H (eds) Encyclopedia of GIS. Springer, New York, pp 235–239
May M, Merten S, Brauner J (2005) Open source für Geodateninfrastrukturen – eine echte Alternative? In: Proceedings CORP 2004, Vienna, 2005
Mische V (2009) Geodata and CouchDB. Presentation at FOSS4G 2009 Conference, Sydney, AU. Available from http://vmx.cx/blog/2009-11-17/geodata-and-couchdb.pdf accessed on 28 May 2010
Rajabifard A, Williamson IP (2001) Spatial data infrastructures: concepts, SDI hierarchy and future directions. In: Proceedings Geomatics, Australia, 2001
Ramsey P (2007) The state of open source GIS. In: FOSS4G 2007 Conference, Vancouver, BC, Canada. Available from www.refractions.net/expertise/whitepapers/ accessed on 28 May 2010
Ramsey P (2009) The OpenGeo Architecture. Available from opengeo.org accessed on 28 May 2010
Reid J, Martin F (2001) The open source movement and its potential in implementing Spatial Data Infrastructures. In: Proceedings international symposium on spatial data infrastructure, Melbourne, Australia. Available from http://www.geo.unizh.ch/publications/ sstein/pubothers/reid_martin_foss_for_sdi_2001.pdf accessed on 28 May 2010
Sanz-Salinas J-C, Montesinos-Lajara M (2009) Current panorama of the FOSS4G ecosystem. Upgrade 10(2):43–51
Smits P (2003) INSPIRE: infrastructure for spatial information in Europe. Presentation held on Sept. 17 2003, open European symposium on GI interoperability, Segovia, 17 Sept 2003
Smits PC, Düren U, Østensen O, et al (2002) INSPIRE Architecture and Standards Position Paper. Available from http://inspire.jrc.ec.europa.eu/index.cfm/pageid/241/documentid/809 accessed on 28 May 2010
Steiniger S, Bocher E (2009) An overview on current free and open source desktop GIS developments. Int J Geograph Inf Sci 23:1345–1370
Steiniger S, Hay GJ (2009) Free and open source geographic information tools for landscape ecology. Ecol Inf 4:183–195
Steiniger S, Weibel R (2009) GIS software – a description in 1,000 words. Available from http://www.geo.unizh.ch/publications/sstein/gissoftware_steiniger2008.pdf accessed on 28 May 2010
Sveen AF (2008) Use of free and open source GIS in commercial firms. Masters thesis, Norwegian University of Technology and Science. Available from spatialserver.net/osgis/ accessed on 28 May 2010

MIX
Papier aus verantwortungsvollen Quellen
Paper from responsible sources
FSC® C105338

If you have any concerns about our products,
you can contact us on
ProductSafety@springernature.com

In case Publisher is established outside the EU,
the EU authorized representative is:
Springer Nature Customer Service Center GmbH
Europaplatz 3, 69115 Heidelberg, Germany

Printed by Libri Plureos GmbH
in Hamburg, Germany